STRUCTURE FORMATION IN ALLOYS

FIZICHESKIE OSNOVY FORMIROVANIYA STRUKTURY SPLAVOV

ФИЗИЧЕСКИЕ ОСНОВЫ ФОРМИРОВАНИЯ СТРУКТУРЫ СПЛАВОВ

STRUCTURE FORMATION IN ALLOYS

by

Il'ya Vasil'evich Salli

Authorized Translation from the Russian

CONSULTANTS BUREAU
NEW YORK
1964

The original Russian text was published for the Dnepropetrovsk
State University by Metallurgizdat (State Scientific and Technical
Press for Ferrous and Nonferrous Metallurgy) in Moscow, in 1963.

Илья Васильевич Салли

ФИЗИЧЕСКИЕ ОСНОВЫ ФОРМИРОВАНИЯ СТРУКТУРЫ СПЛАВОВ

Library of Congress Catalog Card Number 64-20294

© 1964 Consultants Bureau Enterprises, Inc.
227 West 17th St., New York, N.Y. 10011

ISBN-13: 978-1-4684-1562-9   e-ISBN-13: 978-1-4684-1560-5
DOI: 10.1007/ 978-1-4684-1560-5

CONTENTS

PREFACE

Properties of alloys are determined to a considerable extent by the form, the dispersity, the composition, and the quantitative relationships of the structural components.

This monograph is an attempt to present the mechanism of the formation of the structure of alloys from the viewpoint of the most modern theory of phase transformations. The metastable state is treated at length because there are few data in the literature up to the present time.

The book concerns the conditions determining the creation of the different phases in alloys. The formation of crystals of different compositions and dispersities is described. The cause of different degrees of metastability of alloys, the mechanism of the transformation of metastable systems into the stable state, and other problems are analyzed. The most widely used alloys were investigated.

The authors tend to avoid demonstrations based on cumbersome calculations and whenever possible replace them with conceptual models.

Some of the problems described here resulted from discussions during the seminar of the Metal Science Faculty of the Dnepropetrovsk Metallurgical Institute, directed by K. P. Bunin.

The basic experiments were made by the author in the laboratory of the Department of Metal Physics of Dnepropetrovsk State University in collaboration with E. V. Finagin, A. N. Shul'diner, E. Z. Graifer, E.I. Psarev, I. I. Pesetskii, V. L. L'nyanyi and I. S. Miroshnichenko.

The author is grateful to D. S. Kamenetskaya and I. L. Aptekar' for discussions of many fundamental problems.

CHAPTER 1

GENERAL CONCEPTS OF THE STRUCTURE
OF AN ALLOY

Metals and alloys, usually used in the solid state, can be prepared by crystallization from the liquid phase, by the metal-ceramic method, by condensation of vapor, and by electrolytic precipitation.

Each of these methods has its particularities. The first method is the most widely used because the technique is relatively simple and its yield is immense. In this case the components of the alloy are mixed in the liquid state and the structure of the alloy is formed as the liquid solidifies. The second method is based with properties characteristic of the alloy. This method is coming into wider use because of the development of powder metallurgy. The third method consists of the evaporation of one or several metals which are condensed on substrates or machine parts. The yield of this method is very low and it requires expensive equipment. However, it is irreplaceable in many branches of technology and science. The fourth method is peculiar in some respects. Where in the first three methods the decisive factors are the temperature and the composition of the medium from which the alloy is obtained, in electrolytic precipitation the principal factors are the electrical parameters of the process.

The conditions can be varied in each of these methods: the molten solution can be solidified at different cooling rates, in different media, under the effect of ultrasound, under the effect of different fields, etc. In the solid state the alloy can be subjected to different kinds of heat treatment to produce transformations which drastically change the properties of the alloy.

However, no matter what the method or the conditions, the formation of the structure of the alloy is governed by the laws of phase transformation.

Phases are uniform parts of the alloy with distinct physicochemical and thermodynamic properties. One phase is separated from another by a separation boundary. These phases may exist in different states and have different compositions and shapes.

In the final analysis the properties of the alloy are determined by the properties, the shapes, the quantitative relationships, the relative positions, and the dispersities of phases in a given system.

Before studying the relationships governing the formation of the structure of the alloy, let us examine the processes which may occur in the phases themselves in different states and at different temperatures.

1. Gaseous, Liquid, and Solid States

The properties of phases in different states are determined by the kinetic energy of motion and by the potential energy of interaction and relative positions in space of atoms (molecules) of which the phase is composed. Let us examine the atomic structure of a simple body consisting of a large number of identical atoms in different states under constant pressure.

In the gaseous state at relatively high temperature the atoms have high kinetic energy. They move along different trajectories at high velocities. The length of the free path and the frequency of collision of atoms at a given temperature is determined by the number of atoms per unit volume or by the pressure. The distribution of atoms in space is random. The number of collisions between atoms and the length of the free path increase with increasing volume. At high velocities the atoms behave as elastic spheres at the moment of collision.

This is the picture of a so-called ideal gas. In an ideal gas the interactions between atoms play almost no role. In reality, however, there is a force of attraction between any two atoms, and this force of attraction is the result of complex interactions between the forces of repulsion and attraction of the nucleus-electrons system. Without giving a detailed description of the character and types of atomic interactions, which can be found in many monographs and textbooks, let us note only that the forces of atomic interaction (attraction) decrease very rapidly as the distance between atoms increases. The forces of attraction at distances greater than two to three atomic diameters must be taken into account only in the case of very precise calculations of the structure. They become important only at distances close to the diameter of the atom. Thus, in an ideal gas the force of atomic interaction is manifest only at the moment the atoms collide, but because of the high kinetic energy of motion these forces do not affect the distribution of atoms in space. It should be emphasized that at high temperature, which is the measure of the average kinetic energy of the atoms composing the gas, the atoms are being continuously displaced. In this state the motion of the particles is characterized by maximum disorder.

The effects of the force of atomic interaction become significant as the temperature decreases. Some of the atoms unite into groups of two, three, and even more atoms. These unions are very short-lived, but at each moment of time there is always a certain number of such groups. These unions of atoms are favored by the fact that not all the atoms have the same kinetic energy. The kinetic energy of different atoms can differ considerably from the average value because of the random motion. As the result, so-called fluctuations of density occur in the gas. In places where the atoms are assembled in more dense groups the density of the gas is higher than in other places. The magnitudes of these fluctuations may be very different, but all are characterized by a very short "lifetime" and there is no sharp (stable) separation boundary with neighboring areas. It is quite obvious that these peculiarities are somewhat interdependent. A small group of atoms may have a much sharper separation boundary, but then it has a very short lifetime, since the first collision with another atom may disperse this group completely. Large fluctuations have a longer lifetime, but are subject to much more frequent impacts from surrounding atoms, and therefore experimental data on fluctuations are obtained indirectly. The data usually concern the size of the fluctuation and the characteristic relative positions of atoms in them [5-9]. As to the lifetime of fluctuations, there are only theoretical assumptions [10], and there are no data whatever on the structure of the boundary zone. In fact, as will be shown, the principal role in phase transformations is not so much the size of fluctuations as the structure of the boundary zone which separates groups of atoms having a distinct coordination from another part of the system. As the temperature of a gas decreases, the degree of ordering increases, an increasing number of atoms collect into more and more dense groups, and at a certain temperature the gas becomes a liquid.

We shall not dwell here on the mechanism of the formation of a liquid, but proceed immediately to examine the structure of the liquid. The liquid state is characterized by a constant volume and by a boundary of separation from the surrounding medium. The most widely accepted model of the structure of a liquid is the model of thermal motion of the particles developed by Frenkel' [11].

Most of the atoms of a liquid oscillate around equilibrium centers. The time in which the atom is at a given point (settled lifetime) is determined by the relationship:

$$\tau = \tau_0 e^{\frac{Q}{kT}},$$

where τ_0 is the period of oscillation of thermal motion; Q is the activation energy; k is the Boltzmann constant; and T is absolute temperature.

With increasing temperature the exponent decreases very rapidly. Thus, the different atoms in a liquid mix continuously as the result of single displacements from one center of oscillation to another. If we judge the structure of a liquid from x-ray data, which gives an idea of the average statistical distribution of atoms during a time $\tau_1 \gg \tau$, it turns out that in many areas of the liquid there are atomic groups with a certain degree of order.

The experimental results allow us to find not only the first coordination sphere but sometimes even spheres further removed. It is clear that these results are the sum of an enormous number of moments when this relatively high ordering of atoms occurs in different areas of the liquid. The transformation of an ordered group into a disordered state can occur "smoothly" as the result of the transfer of order in a given direction to a neighboring group of atoms.

Therefore it follows that the boundary between ordered and unordered areas of the liquid must be highly indistinct in space and time.

The solid state is characterized by a high degree of ordering; the atoms form a crystal lattice. The amplitude of oscillations of atoms in the lattice also depends on the temperature. As the temperature approaches the melting point, the amplitude of oscillations increases from 0.11 to 0.13a; a is the interatomic distance. In an ideal crystal at a very low temperature the atoms are strictly at the lattice sites, so that the x-ray diagrams reveal clear interference maxima from faces with relatively high indices.

In real crystals there is always a certain degree of disorder, because, aside from a large number of atoms occupying the lattice sites, there are dislocated atoms, vacancies, and other structural defects. The existence of these defects is related to the fluctuation of energy as the result of some randomness in the distribution of thermal motion during transfer of the motion from one atom to another.

In solid, one-component systems the atoms also mix, although in this case the number of atoms moving at any given moment is much smaller than in a liquid.

2. Solutions and Intermediate Phases

The motion of different types of atoms is different in different states because:

1. Since the energy of interaction between atoms of different types and atoms of the same type is different, there is either complete mixing, separation of the system into areas with different ratios of the different atoms, or even complete separation of atoms of a given type.

2. As the result, there is additional disorder from the disruption of the regular alternation or separation of atoms. In the gaseous state all metal alloys form mixtures with infinite solubility. In the liquid state most of the components of alloys mix infinitely, particularly at high temperatures. However, in a number of systems (Pb–Zn, Pb–Cu, Al–Na, etc.) the forces of interaction between the same atoms become so great with decreasing temperature that separation begins in the liquid state. The atomic interaction is even greater in the solid state. In the solid state there are solutions with unlimited solubility, primary solutions with limited solubility (which varies with temperature), and finally a practically complete insolubility of the components of the system. In first approximation the miscibility of components is characterized by the energy of mixing per atom of the alloy:

$$V' = V'_{AB} - \frac{V'_A + V'_B}{2},$$

where V'_{AB} is the energy per atom in the alloy and V'_A and V'_B are the same energies in the initial components.

An ideal solution is formed when $V' = 0$. In this case the distribution of atoms of different kinds in the lattice sites resembles disorder in an ideal gas if one considers only alternation of atoms of different types. If $V' > 0$ then the binding energy between different types of particles is greater than between particles of the same type. The forces of attraction are in opposite relationship, and consequently the atoms of the A type are mostly surrounded with atoms of the same type, and the same is true of B-type atoms.

It is quite clear that the thermal motion resulting from the temperature of the system acts in the opposite direction: it favors complete mixing. The relationship between these two factors determines the limit of solubility of the primary solid solution.

The magnitude of the mixing energy given by the equation above is only a first approximation. In reality, this value depends on the temperature and the concentrations of the components. During melting of the components the mixing energy is determined not only by the binding forces but also by internal deformations. In the final analysis the mixing energy depends on the difference between the diameters of atoms, their valences, the peculiarities of the crystal lattices of the initial components, etc. Continuous solid solutions of two components can form only under the following conditions:

1. The atomic radii of the components must not differ by more than 8% (I. I. Kornilov [293]) or 15% (Hume-Rothery [2]).

2. The components must have the same valence.

3. The crystal lattices of the components must be isomorphic. These conditions are necessary but not sufficient for the formation of a continuous series of solid solutions between two components. For instance, copper and silver melted together do not form a continuous series of solid solutions in spite of the fact that all three conditions are met.

During the formation of solutions not only the density fluctuates but also the concentrations. Fluctuations of concentrations are areas of the solution in which the composition differs greatly from the average concentration. A great deal of experimental work has indicated that the concentrations can differ greatly from the average concentration, even up to complete separation of the components [12-18]. Some authors have found fluctuations of concentrations in liquid solutions. For example, in liquid cast iron the size of the areas consisting almost entirely of carbon is 10^{-6} cm [19]. The greatest fluctuation of concentrations occurs in solutions with positive mixing energy ($V' > 0$). However, the possibility of fluctuating concentrations in almost ideal solutions is not excluded.* On the basis of general considerations of the character of thermal motion of atoms in liquids, one can conclude that:

1. Fluctuations of concentrations have no sharp boundary of separation with the surrounding liquid. Regardless of the size of the area in which the concentration differs from the average, the concentration always varies smoothly over a distance of the order of the size of the fluctuation itself.

2. The "lifetime" of such fluctuations is very short. It can be represented rather as a wave which is displaced in the liquid so that the concentration decreases on one side of the wave and increases on the other. At the moment of maximum separation of atoms of the components there may occur a structure which is close to the structure of the components in the solid state. This fluctuation resembles a liquid phase, since it does not have a definite shape or boundary separating it from other parts of the solution. However, it also has the characteristics of one of the components in the solid state, since (at least for a short period of time) it has a structure closely resembling the lattice of this component as the result of the forces of atomic interaction.

When atoms of elements with quite different physical and chemical properties are mixed together, intermediate phases are formed in the solid state. These intermediate phases are phases with a normal valence, penetration phases, electronic compounds, σ-phases, etc.

In molten alloys groups of atoms with a composition close to that of the intermediate phase may occur even if the average composition of the alloy is very different from the composition of the intermediate phase, and the atoms in these groups form an order resembling the order in the lattice of the intermediate phase [20].

However, the main properties of these groups of atoms remain unchanged. The situation changes drastically with increasing temperature. At high temperatures there are almost no fluctuations of the concentration; with decreasing temperature they become more stable. The fluctuations are the result of a complex interaction between such factors as the random character of thermal motion in a system composed of a very large number of particles and the interaction between separate atoms of the system.

During transformation into the solid state the effect of thermal motion decreases, and as the result the distribution of atoms becomes more orderly and the probability of the occurrence of fluctuations of the concentration decreases again. However, at very high temperatures there is random motion of some atoms even in the solid state. Therefore, atoms of one kind can gather in different areas of the solid solution and form local fluctuations of the concentration. The general character of these fluctuations resembles that of fluctuations in a liquid, the only difference being that in a solid their mobility is much lower.

3. Shape of Structural Components, Defects, and Grain Boundaries

Examination of the structure of metals and alloys under a microscope shows that the shapes of crystallites composing an alloy can be divided into several different types (Fig. 1): 1) polyhedral crystals with plane boundaries; 2) oval and spheroidal crystals; 3) plates and needles; 4) branched crystals (dendrites, spherulites, and split crystallites).

_____

* For more details on the occurrence and properties of fluctuations of concentrations see M. I. Shakhparonov [294, 295].

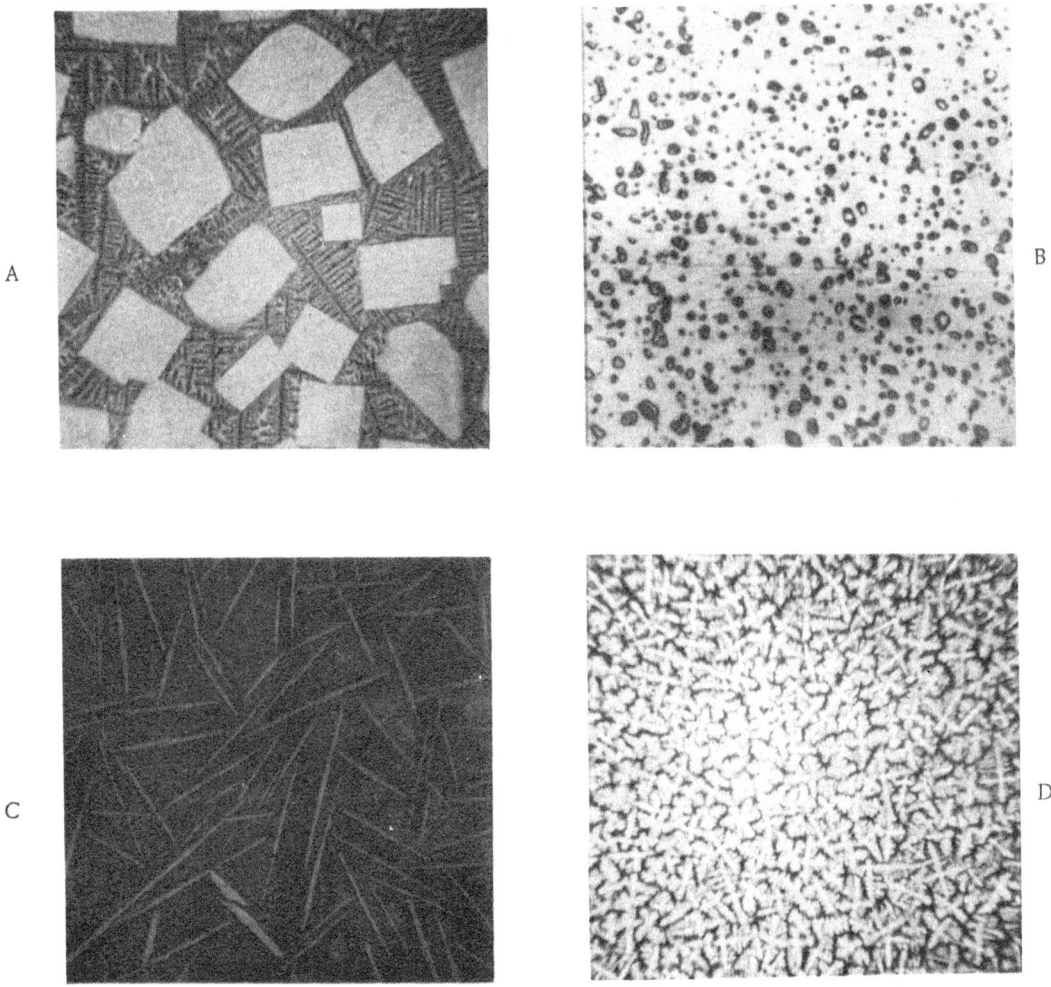

Fig. 1. Shapes of crystals. A) Polyhedrons with plane faces; B) oval; C) plates; D) dendrites.

Also, many alloys consisting of a mixture of different phases contain colonies of regularly alternating plates of a given shape (pearlite colonies) or consisting of parallel threads of one phase separated by a second phase (ledeburite colonies). Sometimes the colonies have a ray structure similar to spherulites (Fig. 2).

The shape of the structural components depends to a great extent on the type and conditions of phase transformations. The conditions of phase transformations also determine finer peculiarities of the structure of separate crystallites. Among these peculiarities are the microheterogeneity in the composition of crystallites and different types of defects and imperfections. As the result of intracrystalline liquation, the whole crystallite looks as if it were a stabilized fluctuation. It begins to split up and disappear as the thermal motion of the atoms increases. In some cases this leads to considerable changes in the properties of the whole alloy. Usually, intracrystalline liquation can be revealed by special etching agents, radioactive tracers, local analysis, x-ray analysis, and measurements of microhardness.

Electron microscopic and x-ray studies show a number of imperfections in crystals. These imperfections play a very important role in phase transformations, recrystallization, deformation, and other processes.

A great number of studies and reviews have been made of the superfine submicroscopic structure of alloys [1, 2, 3, 21-23, 81].

Real crystallites usually have defects of different sizes. Small defects (of the order of the atoms) are called point or atomic defects: vacancies, dislocated atoms, impurity atoms in lattice sites and intersites, and also pile-ups of such impurities and vacancies.

Fig. 2. Raylike structure.

The second type of defect is linear defects, among which dislocations are the most important. These defects are small in two dimensions with a much extended third dimension. These defects can be divided into linear and screw dislocations. Linear dislocations can be represented as defects formed as the result of penetration of extraplanes or halfplanes into the crystal lattice. In Fig. 3a a positive dislocation is indicated by the sign ($\perp$) and the negative by ($\top$). The edges of halfplanes, which may be either a straight line, a spiral, a closed loop, etc., are called lines of dislocations. A screw dislocation is shown in Fig. 3b.

The third type of defect consists of boundaries between crystallites, blocks, twins, and phases.

Point defects occur when not all the atoms of the system have the same energy. Among the large number of atoms there are always some which have a high energy (E_1) sufficient for the atom to leave the lattice site and move a certain distance, finally settling between sites. And, of course, there are atoms with energies lower than the average energy (E). The number of atoms with energies necessary to form defects is determined by the Boltzmann equation

$$n = n_0 e^{-\frac{\Delta E}{kT}},$$

where n_0 is the total number of atoms and $\Delta E = E_1 - E$ is the activation energy.

The number of dislocated atoms increases with increasing temperature; consequently, the number of vacancies also increases. Vacancies can also migrate into the metal from the surface as the result of evaporation of some surface atoms [11]. At each temperature there is an equilibrium number of point defects which is determined by the average kinetic and potential energies of atoms. The process of recombination of vacancies and dislocated atoms plays an important role in establishing equilibrium. When these two defects meet they annihilate each other. The probability of such a meeting is proportional to the number of defects of both types and depends on the temperature of the alloy.

Dislocations are unstable defects. They usually occur during phase transformations or plastic deformation. Their number decreases as the result of prolonged annealing. Dislocations move easily as the result of stress,

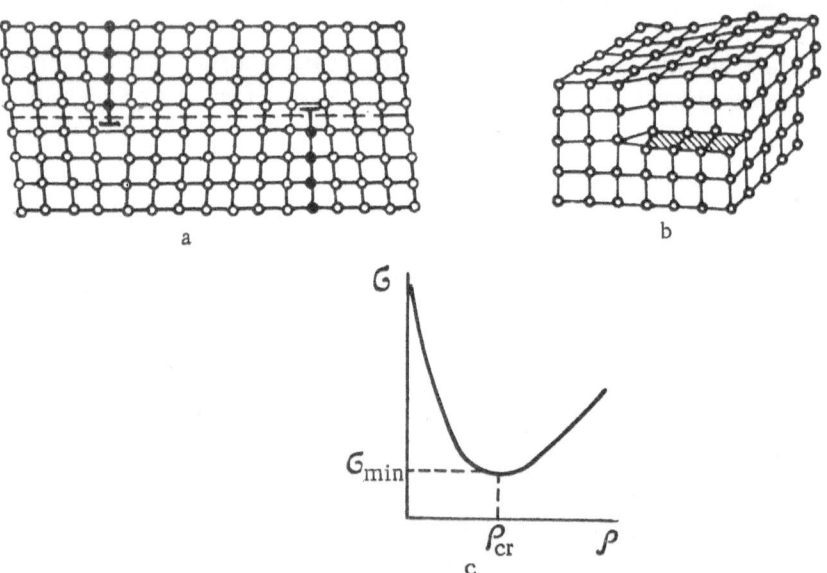

Fig. 3. Dislocations. a) Linear; b) screw; c) variation of the resistance to shear σ with the density of dislocations ρ.

6

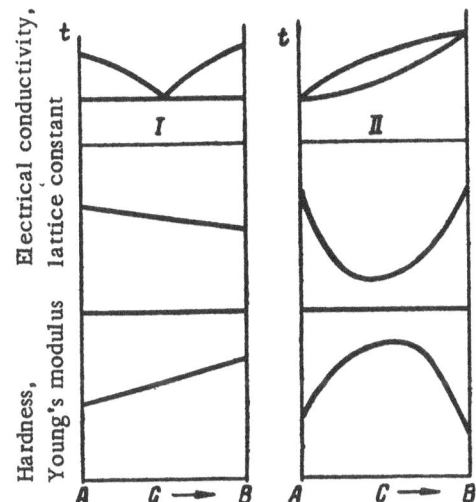

Fig. 4. Variation of the properties of alloys with the composition for two types of interactions between components. I) Mixture of two phases; II) continuous series of solid solutions.

while linear dislocations can move as the result of thermal motion. Dislocations of different signs attract each other and dislocations of the same sign repel each other. Point defects and other lattice defects – and especially grain boundaries – prevent the motion of dislocations.

The boundary between crystallites is one of the most important structural elements in metals and alloys.

In general terms, the boundary is a layer of atoms adjacent to the surface of contact between crystallites or between crystallites and other phases. The boundary atoms have higher kinetic and potential energies than the atoms within the crystal. The forces of interaction of the atoms with their neighbors are compensated within the crystal and the interaction energy has a relative minimum. The atoms at the boundary with a vacuum or saturated vapor or liquid are held by a smaller number of neighbors, and consequently their bonds with the crystal are much weaker. Uncompensated bonds provide a certain excess of potential energy which is called surface energy. This energy per unit surface is surface tension. The number of uncompensated bonds at the boundaries of crystallites of a single-phase system depends on the relative orientations of the crystallites. If crystallites of different phases are adjacent to each other then the value of the surface energy also depends on the difference between the energies of atomic interactions of the phases in contact.

At the present time the structure of boundaries is not well known.

In a single-phase system the boundary zone is apparently saturated with dislocations when the disorientation is not great – when the angle between mosaic blocks, twin boundaries, and crystallites is no greater than 20° When the disorientation is greater the boundary also has a considerable number of vacancies.

Some investigators consider that the grain boundaries (in a single-phase system) have a lower melting point than the metal. It was shown in [286-288] that the melting point at the boundary can by 0.1-4°C lower than the melting point of the crystal itself. At the same time, the melting point of the boundary between twins is no lower than the melting point of the metal itself. The nature of this phenomenon is not yet clear, although it can be attributed to the enrichment of the boundary layer in impurities always present in the metal. The boundaries are of primary importance in the formation of the structure of alloys because the transformation of one phase into another begins with the formation or destruction of the boundaries between phases. The concept of the separation boundary is intimately related to the concept of phases and states.

4. Structure and Properties of Alloys

One of the most important and yet most difficult problems of metal science and the physics of metals is to determine the relationship between the structure and the properties of alloys.

At the present time a great effort is being made by theoretical and experimental scientists to relate the structure of the strength and ductility of single-phase systems. Some progress has been made as the result of the dislocation theory, the theory of plastic deformation [26-29], and also the theory of the strength of materials based on the effect of cracks [30-32, 284]. The resistance of metals to rupture is determined to a certain extent by the simultaneous effect of both dislocations and cracks. Many authors have reviewed the results in this area of investigation [1, 21-23, 33]. They have attached particular importance to different types of defects and also to the submicroscopic and atomic structure of alloys.

The resistance of metals to deformation depends on the density and mobility of dislocations. The mechanism of plastic deformation consists in the directed motion of dislocations under the effect of external forces. Each dislocation which comes out to the surface of the crystal leads to the elementary slip of one plane of the crystal lattice with respect to another. An increase of dislocation density up to a certain limit increases the plasticity of metals. However, an increase in dislocation density beyond this limit strengthens the metal.

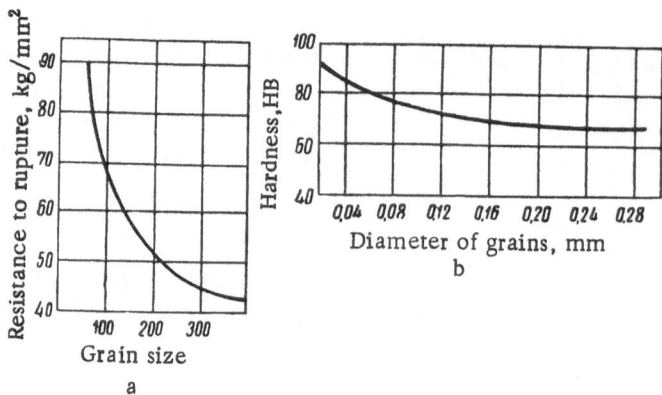

Fig. 5. Effect of the ferrite grain size on: a) the resistance to rupture
[35]; b) hardness [36].

The strengthening of metals is the result of interactions between dislocations and the creation of disloca-tion walls, "clouds," and other barriers to the motion of dislocations. The effect of dislocation density on the strength of the metal can be represented by the curve shown in Fig. 3c [283]. This figure indicates that resist-ance to slip σ has a maximum corresponding to the critical density of dislocations ρ_{cr}. For annealed metals the dislocation density reaches 10^6-10^8 dislocations per cm^2 and for metals subjected to a high degree of cold working it reaches 10^{11}-10^{12} dislocations per cm^2.

Thus, the strength of a metal can be increased by an increase or decrease of dislocation density. The strength of a metal can also be increased by creating different barriers to the motion of dislocations – by adding impurities with low mobility or by subjecting the metal to treatments resulting in polygonization.

Dissolved atoms are concentrated in the immediate vicinity of dislocations. The atoms which stretch the lattice are situated inside the dislocations. The dislocations are blocked by impurity atoms and can detach themselves from them only under the effect of high stress. Only small amounts of impurities are required to block the dislocations.

Strengthening due to polygonization results from the formation of walls of dislocations of the same sign.

Strengthening of metals by elimination of dislocations is a method with great promise but it is not yet perfected. At the present time, very strong threadlike crystals (whiskers) 2-10 μ thick can be grown. Thus, crys-tals of pure iron have an ultimate strength of ~1400 kg/mm^2 and those of copper ~700 kg/mm^2.

The theory of the strength and ductility of single-phase systems explains and even predicts the tremendous number of experimental facts and phenomena occurring in single crystals during deformation. The situation is much more complicated in the case of alloys with a complex internal structure.

Qualitative relationships between the properties and the composition of alloys with a complex internal structure have been determined by N. S. Kurnakov [25, 34, 296]. He relates the properties of the alloys to the shape of the state diagram (Fig 4). The relationship shown in this figure between the composition of the alloy and the properties holds good only when the structural components have a relatively simple shape, are not too small, and the tests are made at normal temperatures.

However, many cases are known today where as the result of a change in the structure alone (the phase composition remaining the same) the properties of the alloy change greatly. *

For example, by changing the grain size one can change the resistance to rupture and the hardness. Figure 5 shows the effect of the grain size on the hardness of ferrite [35]. In [36] it was found that the relation-ship between the grain size and the temperature at which the brittle state becomes a plastic state – determined by impact tests – is linear in iron containing 0.02% C (the grain size changes from 16 to 512 grains per mm^2).

---

* The role of structure was noted for the first time in [296] and then analyzed in greater detail in [297-299].

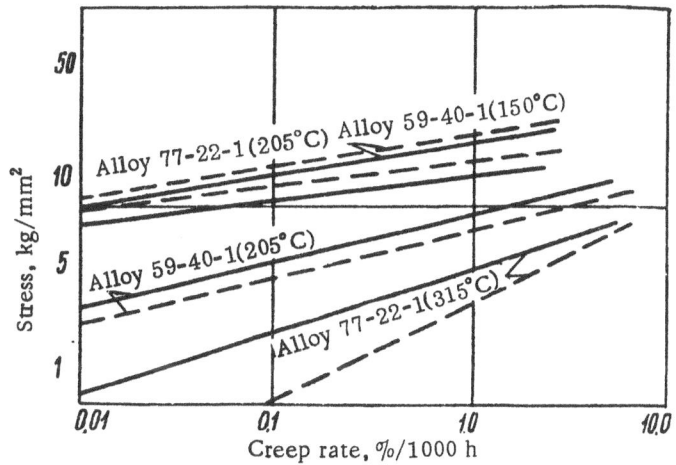

Fig. 6. Effect of grain size on the creep rate. ——) Large-grain
alloys; — — —) small-grain alloys [37].

TABLE 1

Concentration of Zn, %	Relative elongation of alloys, δ, %	
	cast	rolled and annealed
10	~36	~38
20	~45	~50
30	~57	~63
40	~35	~50
50	~ 5	~25

The grain size has a particular effect on creep. It is well known that at high temperatures viscous flow begins along the grain boundaries, while at low temperatures the boundary is stronger than the crystal itself. Creep, which depends on grain size, follows the same pattern. The difference in creep of large and small grain samples of hot rolled Cu–Zn–Sn alloys was investigated in [37].

The results of this investigation are shown in Fig. 6. It can be seen that alloys with small grains have higher resistance to creep at low temperatures, while alloys with large grains have a higher resistance to creep at high temperatures.

Intracrystalline liquation has a great effect on the properties of single-phase alloys. For example, the difference in the properties of cast and annealed brass [38, 39] are given in Table 1, where it can be seen that heterogeneities in the solid solution effect the ductility.

The effect of intracrystalline liquation on the properties of single-phase alloys was also noted in [301,302]. A considerable amount of work on the effect of dendritic liquation on the properties of steel was done in [153].

In systems consisting of several phases the size and shape of the structural components have a great effect on the properties of the alloy.

Second-order heterogeneities or microheterogeneities of crystallites in multiphase alloys also have a great effect on the properties [299, 300].

The phase transformations and the size and shape of the second phase can be changed as the result of coalescence and spheroidization. Both processes decrease the elastic limit and increase the ductility of the alloy.

S. Z. Bokshtein [40] has done considerable work in this area. He studied the variation of the mechanical properties of alloyed steel as a function of the size of their carbides. He showed that the hardness of steels tempered at high temperatures (other conditions being equal) is determined by the total surface of the carbides and can be expressed by the following equation

$$H = H_0 + kS,$$

where H_0 is the hardness of ferrite and k is the strengthening coefficient of steel resulting from the dispersity of

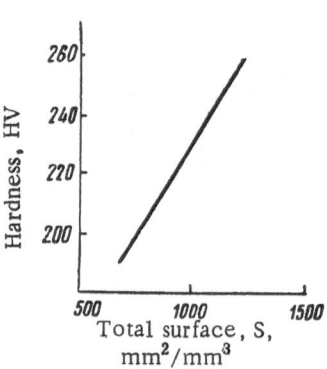

Fig. 7. Variation of hardness
with the total surface of carbides in carbon steel [40].

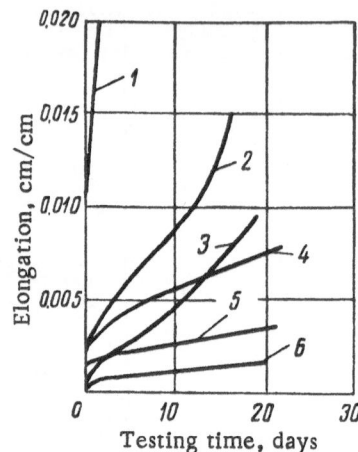

Fig. 8. Effect of heat treatment on creep in steel
containing 0.11% C and 0.52% Mo under a stress
of 14.2 kg/mm^2 and a temperature of 550 C [44].
1) Spheroidized; 2) as delivered; 3) quenched
in water and tempered; 4) cooled in water; 5)
strengthened; 6) quenched in oil and tempered.

carbides, which depends on the value of the surface tension at the ferrite-carbide boundary (determined empirically);

$$S = \frac{6V_\varepsilon}{\alpha_{cp}},$$

where V_ε is the total volume of carbides and α_{av} is the arithmetic average diameter of the carbide particles.

Figure 7 shows the variation of the hardness of carbon steel with the total surface of the carbides.

In steel containing layered pearlite the hardness also depends on the dispersity (the average distance between the plates of the carbide component [41]). In steel with a layered structure the yield point and hardness are higher than in steel with a globular structure, the amount of carbon being the same.

It is well known that phases consisting of small particles (grain size less than 10^{-5} cm) strengthen the mother solid solution. However, the distribution of the particles is an important factor. When the particles are uniformly distributed throughout the whole crystal the alloy is strong and also resilient. If, however, the particles are concentrated essentially at the boundaries between the grains of the mother phase the alloy is very often brittle [42, 43].

The same phenomenon occurs when the particles surround the crystallites of the second phase. Alloys with a structure of the Widmanstatten type also have a low resilience.

The shape of the structural components also affects creep. Thus, the dispersed phase has a strengthening effect (as determined by creep tests) only if this phase is resistant to coalescence and spheroidization. If it is not resistant then the resistance of the metal to creep may be very low. There are numerous data indicating that the spheroidal dispersed phase has less creep resistance than any other shape of this phase. Figure 8 shows the resistance to creep of steels with different structures [44].

The properties of alloys in a nonequilibrium state can easily be changed. Among the alloys in the non-equilibrium state are alloys containing supersaturated solid solutions or metastable intermediate phases. Alloys containing supersaturated solid solutions can be made very hard and strong simply by quenching. In most cases their mechanical properties are improved by aging. The variation of the properties of alloys with the type of heat treatment have been described in many publications, among which the most important are those by Shteinberg, Kurdyumov, Konobeevskii, and their students.

Among the alloys containing metastable intermediate phases are cast irons and alloyed steels containing several types of carbides. It is well known that the changes in the structure from white to forgeable cast iron are the causes of the great difference in the strength and other properties [45].

Unfortunately, all these data do not elucidate a quantitative or even a qualitative relationship between the metallographic structure and the properties of alloys. No quantitative relationship has been found between the structure and any specific properties of alloys. To find such a relationship it is first necessary to establish the general laws of the formation of the structure of alloys.

CHAPTER 2

SOME CONCEPTS OF THERMODYNAMICS

Thermodynamics concerns the transformation of one form of energy into another, and therefore it is used in many different branches of science and industry. From a few basic laws thermodynamics gives us precise relationships between the different parameters determining the state of a system under given conditions. At the present time, thermodynamics can be divided into two parts: the thermodynamics of reversible processes and the thermodynamics of irreversible processes. The thermodynamics of reversible processes is usually used to study equilibrium conditions between different phases of a system, to analyze phase diagrams, and to calculate calorimetric data for different reactions. The thermodynamics of reversible processes also indicates clearly the direction of the processes without specifying the mechanism or the rate of transformation into a new state.

Irreversible processes include heat transfer, diffusion, electrical conductivity, internal friction, thermodiffusion, etc. In other words, it describes phenomena which accompany the transformation of a system from one to another state. The thermodynamics of irreversible processes is nothing but a generalization of classical thermodynamics and laws determined experimentally by the study of transition states. The main problem of the thermodynamics of irreversible processes concerns the determination of the rate of transformation of a system from one state to another.

The thermodynamics of reversible as well as irreversible processes uses parameters resulting from the motion and interaction of the great number of particles composing a system.

Among these parameters are temperature, pressure, volume, concentration, diffusion coefficient, heat conductivity, electrical conductivity, etc. Some of the simplest relationships between these parameters, which have been experimentally confirmed, are the basis for the derivation of relationships for more complex cases. The molecular mechanism of processes is studied by statistical physics, which originated from the kinetic theory of gases created by Lomonosov, Clausius, Maxwell, and Boltzmann. At the present time, statistical physics gives a strict description of all thermodynamic relationships on the basis of the statistical relationships obeyed by the phenomena occurring in a system consisting of a very large number of particles. Statistical physics describes the physical meaning of such magnitudes as temperature, pressure, and different kinetic coefficients such as diffusion, heat conductivity, etc.

We shall not dwell here on the strict derivations upon which the physical meanings of some concepts of thermodynamics are based [46, 47]. We shall use only their final results and we shall consider the meanings of only those concepts necessary for the study of the laws governing the formation of the structure of alloys.

1. Internal Energy and Entropy

As we have said, a system consisting of a large number of particles possesses a definite energy which is the sum of the potential and kinetic energies of the particles. Not all the particles have the same energy, but certain average values of the kinetic and potential energies can be ascribed to these particles. The sum of these energies represents the internal energy of the system E. Thus, the internal energy includes all the types of potential energy resulting from the respective positions of molecules and all the types of kinetic energy related to the masses and velocities of the molecules (including oscillations). The internal energy of a system can be used to do a certain amount of work A (e.g., the motion of a piston in a cylinder) or can be used to heat some colder body.

The transfer of heat from a hot body to a cold body is accompanied by scattering of part of the heat because of the randomness in the transfer of energy from atom to atom. It is quite characteristic that, all other

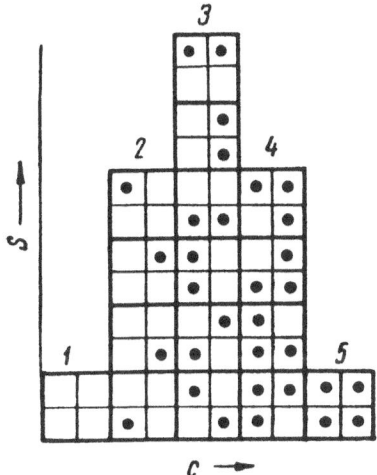

Fig. 9. Number of distributions.

conditions being equal, the amount of degraded energy is determined by the type and velocity of the motion of the particles composing a body, and therefore it is directly connected to the structure of the body and the temperature at which the process occurs.

The measure of degraded internal energy is the amount of scattered heat (dQ) divided by the temperature at which the measurements are made:

$$dS = \frac{dQ}{T}.$$

The value of S is called entropy:

$$S = \int\limits_{0}^{T} \frac{dQ}{T}.$$

Entropy is the measure of the degraded internal energy per degree of absolute temperature.

In statistical physics this magnitude is defined in a somewhat different way.

At any given temperature and any given value of internal energy some part of the internal energy is spent on internal displacements which do not participate in the transformation of heat into work. If atoms possessed only kinetic energy they would be distributed in space quite at random and the entropy of the system would be maximum. With decreasing temperature, the energy of atomic interactions forces the atoms to acquire first a close and then a far order, and thus also limits the value of entropy. Therefore, at each given temperature the value of entropy can be correlated to the number of possible different distributions of particles in a system W for which the value of internal energy remains constant. W is also called the thermodynamic probability of a state.

Let us take as an example a system consisting of four cells. Let us assume that only one atom can exist in a cell (Fig. 9). Then, in a system consisting of four cells and one atom there can be four distributions. In a system of two atoms there will be six distributions; in the case of three atoms there will again be four; and if the number of atoms and cells is the same there is only one distribution (we consider the case where all the atoms are identical). The number of possible distributions will be given by the equation

$$W = \frac{N!}{N_1! \, (N - N_1)!},$$

where N is the number of cells and N_1 is the number of atoms.

An identical calculation can be made for a solid solution when the cells are occupied by atoms of different kinds or when the system contains vacancies.

The calculation shows that entropy is related to the thermodynamic probability by the following relationship

$$S = k \ln W,$$

where k is the Boltzmann constant.

The type of relationship between the entropy and the concentration can be seen even from the example of the cells and atoms. When all the lattice sites are occupied by atoms then $(N - N_1)! = 1$ and when $N = N_1$ then S = 0. The lattice is in this state at absolute zero.

If we represent the concentration of the solution C as a ratio between the number of atoms of one kind and the total number of atoms or sites, i.e., $C = N_1/N$, and dispose of factorials, using the Sterling approximation,* we obtain the expression

$$S = - Nk \, [C \ln C + (1 - C) \ln (1 - C)]$$

---

* $\ln x! \approx x \ln x - x$. This formula is valid under the condition that $x \gg 1$.

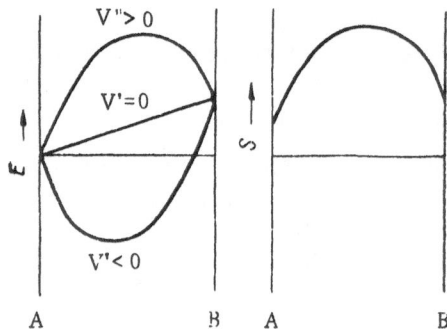

Fig. 10. Variation of internal energy
and entropy.

for the entropy of the solution, which is determined only by the exchange of places of atoms of different kinds. This is the so-called positional entropy.

The total entropy of the solution also depends on the order in which the atoms oscillate. The degree of disorder and the distribution of oscillations between atoms is calculated by a more complicated method, but the calculations show that in first approximation the entropy of the oscillations of atoms in the solution can be taken as equal to the entropy of oscillations of pure components. Then the general expression for the entropy of the solution will be

$$S = [(1-C) S_A + CS_B] - kN [C \ln C + (1-C) \ln (1-C)], \qquad (2.1)$$

where the first bracket is the entropy of oscillations of atoms of the solution.

Let us now find what the internal energy of the solution is equal to. Since the internal energy is determined by the values of the kinetic and potential energies of the atoms of the alloy, and the kinetic energy of the atoms in the alloy is essentially determined by the temperature, the problem is reduced to the calculation of the interactions of different pairs of atoms in the alloy. If we consider only the interaction between closest neighbors, then the following expression

$$U = N_{AA}U_{AA} + N_{BB}U_{BB} + N_{AB}U_{AB} \qquad (2.2)$$

gives the potential energy of the interaction in a binary solution. In this expression N_{AA}, N_{BB}, and N_{AB} are the numbers of the corresponding pairs of atoms and U_{AA}, U_{BB}, and U_{AB} are their interaction energies. When the distribution of atoms is random, the number of atoms of different kinds in contact with a given atom is proportional to C and (1 − C), while the number of different kinds of pairs is proportional to the concentration $C = N_1/N$

$$N_{AA} = \frac{1}{2} NC^2; \quad N_{BB} = \frac{1}{2} N (1-C)^2;$$

$$N_{AB} = NC (1-C).$$

Then, substituting into (2.2), we obtain

$$U = \frac{1}{2} NC^2 U_{AA} + \frac{1}{2} N (1-C)^2 U_{BB} + NC (1-C) U_{AB}.$$

Opening the parentheses and grouping the terms of the equation, we obtain

$$U = \frac{1}{2} N [CU_{AA} + (1-C) U_{BB} + 2C (1-C) \left(U_{AB} - \frac{U_{AA} + U_{BB}}{2} \right)], \qquad (2.3)$$

where $U_{AB} - \dfrac{U_{AA} + U_{BB}}{2}$ is the mixing energy of the solution.

Finally, the internal energy of the solution is determined by

$$E = N [C (1-C)v' + (1-C) U_A + Cv'_B] + K, \qquad (2.4)$$

where K is the kinetic energy of the atoms. This magnitude does not play a very important role in the formation of the structure of alloys because for a given substance in different states it is determined only by the temperature.

Depending on the sign and the magnitude of the mixing energy, the alloying of two components can lead either to the decrease of the internal energy of the alloy or its increase. Whether there will be complete mixing or limited solubility of the components can be determined only be measuring the changes in the entropy and the internal energy simultaneously. Figure 10 shows the dependence of these values on the concentration of the alloy.

14

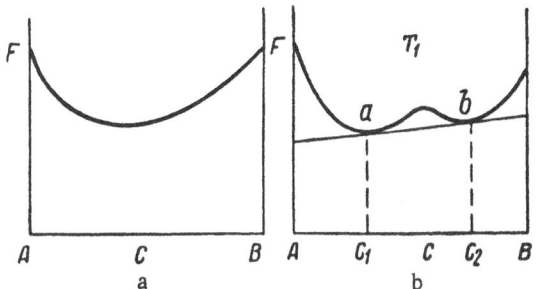

Fig. 11. Variation of the free energy with the concentration. a) V' < 2kT; b) V' > 2kT.

If work F is done on the system it is spent on the increase of the internal energy and part of it is dissipated in the form of heat. According to the law of conservation of energy, we have F = E – Q. For processes occurring at constant temperature and pressure, Q = ST, and consequently

$$F = E - TS, \qquad (2.5)$$

where F is a thermodynamic function which is called free energy.

In a system containing two (or more) components the free energy is also a function of concentration.

If we divide Eqs. (2.1) and (2.4) by N and substitute the results into (2.5) we obtain the free energy of a binary alloy per atom of the alloy

$$F = C(1 - C)V' + (1 - C)V'_A + CV'_B + kT[(1 - C)\ln(1 - C) \\ + C\ln C] - \frac{T}{N}[(1 - C)S_A + CS_B]. \qquad (2.6)$$

A strict analysis of the curve representing the dependence of the free energy on the concentration at different temperatures was made by Kamenetskaya [55]. This analysis leads to the following conclusions.

When V < 2kT the curve has a single minimum (Fig. 11a); if V > 2kT the curve has two minimums (Fig. 11b).

With increasing temperature the curve is displaced downwards and its radius of curvature decreases.* In F and T_c coordinates the curve has the shape of a gutter. If the mixing energy is greater than 2kT, then a convexity occurs at the bottom of this gutter, and this convexity increases with decreasing temperature.

Other conditions being equal, the higher the mixing energy V the greater the radius of curvature. All these consequences, which result from the analysis of Eq. (2.6), are correct only if the mixing energy is independent of temperature.

As we have said, Eq. (2.4) determines the relationship between the internal energy and the concentration only in first approximation. In reality, the mixing energy V in this expression depends on the temperature, at least in the case of the solid solution [48, 49].

The partial derivative of free energy with respect to the concentration $\mu = (\partial F/\partial C)_{T,V}$ (Fig. 12) plays an important role in thermodynamic analysis of phase transformations. In the case of an alloy it indicates the magnitude of the change of the free energy for an infinitely small change in the concentration of an alloy of a given composition (C_1).

The chemical potential of the alloy is represented geometrically by the tangent of the slope of the tangent φ to the curve F(C) for a given composition of the alloy C; since C varies from 0 to 1, we have $\partial F/\partial C = \mu_A - \mu_B$.

---

* The radius of curvature is determined by the equation

$$\rho = \frac{\left\{ \sqrt{1 + \left[(1 - 2C)V' + kTC_T + kT\ln\frac{C}{1-C}\right]^2} \right\}^3}{\dfrac{kT}{C(1-C)} - 2V'},$$

where

$$C_T = \ln\frac{1-C}{C} - (1 - 2C)\frac{V'}{kT}.$$

15

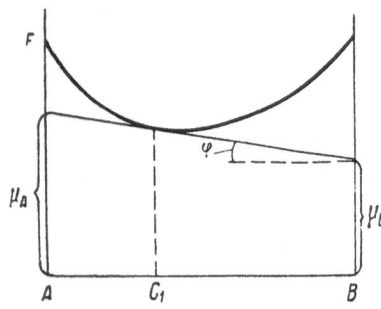

Fig. 12. Chemical potentials.

The partial values of the chemical potentials of the components of the alloys (μ_A and μ_B) are no less important. These values determine the amount of free energy per atom of a given component in the alloy of a given composition at a given temperature. The values of the chemical potentials of atoms of components A and B are indicated by brackets in Fig. 12.

3. Thermodynamic Activity and the Activity Coefficient

From the viewpoint of thermodynamics, ideal solutions are solutions in which mixing of the components does not change the volume, the heat of mixing is equal to zero, and the vapor pressure of the components is proportional to their concentrations. Such solutions obey Raoult's law, and the variation of the free energy resulting from mixing of one mole of a component with an infinitely large amount of solvent is determined by the relationship

$$\Delta\mu_1 = RT \ln \frac{p_1}{p_2}.$$

Here p_1 is the vapor pressure of the dissolved substance above the solution; p_2 is the vapor pressure of the solvent at the same temperature. The ratio $a = p_1/p_2$ is called the activity of a given substance in the solution.

In an ideal solution the activity is equal to the concentration of the substance. Therefore, in solutions close to ideal solutions we have

$$\Delta\mu_1 = RT \ln a_1 \approx RT \ln C_1.$$

In most cases dilute solutions are not very different from ideal solutions. When there is considerable deviation from Raoult's law one introduces the activity coefficient

$$f_1 = \frac{a}{C_1}.$$

In first approximation the activity coefficient and the mixing energy are related by

$$\ln f_1 = -(1 - C_1)^2 \frac{z}{kT} V',$$

where z is the coordination number. The sign and the value of the activity coefficient determine the degree to which the solution differs from an ideal solution.

Thus, we can determine the activity coefficient experimentally, and then determine with great precision the dependence of the thermodynamic magnitudes on the concentration and temperature.

4. Surface Energy

The formation of any surface requires an expenditure of work. If we consider a small reversible change of the surface dS then the work will be expressed as dF = σdS.

The value of σ determined in this way is called the coefficient of surface tension. In other words, this is the amount of work necessary to form a unit surface of separation.

A surface of separation can be formed either in a liquid or in a solid phase. The type of the medium at the boundary of which the surface of a given system is formed or increased is of great importance. The surface tension of a single-component system varies in the following sequence: it is maximum at a solid-vacuum boundary; it is less at a liquid-saturated vapor boundary; and it is minimum at a liquid-solid boundary.

At a vacuum boundary the surface tension is determined by

$$\sigma = \frac{1}{2} \frac{U_0}{S} = \frac{U_0}{2N'r^2\gamma},$$

where U_0 is the energy of interaction between atoms in a given plane of separation; S is the surface of the newly formed boundary; N' is the number of atoms on the surface; r is the radius of the atoms; and γ is the stacking coefficient.

TABLE 2

Metal	σ_{12} theor	σ_{12} exp	Metal	σ_{12} theor	σ_{12} exp
Ag	119	126	Li	25.7	30
Au	145	132	Na	15.7	20
Cu	170	177	K	10.2	–
Pb	35.4	33	Zn	69	–
Al	95	93	Cd	50.4	–
Pt	288	240	Mg	47.3	–
β-Ni	223	255	δ-Fc	193	204
β-Co	217	234			

The surface tension at the phase boundary is

$$\sigma_{12} = \sigma_{11} + \sigma_{22} - R_{12},$$

where σ_{11} and σ_{22} are the surface tensions of the phases at the boundaries with a vacuum; and R_{12} is the interaction energy of atoms at the phase separation boundary relative to the surface of separation. This value is called the work of adhesion.

It is rather difficult to calculate the surface tension because the interaction energy of atoms at the separation boundary is determined by a great number of factors. Up to now, Zadumkin [50] has taken the greatest number of these factors into account in calculating the surface tension.

The equation derived by Zadumkin gives results which agree best with the most precise experimental results. Since the formation of the structure of an alloy depends to a great degree on the value of surface tension, we give some data concerning the surface tension at the metal-melt boundary (Table 2).

To calculate the surface tensions by the equation derived in [50] it is necessary to known the heat of sublimation and the heat of fusion, the close order in the melt, and the orientation factor.

There is a series of formulas with which one can determine the surface tension. Becker [51] gives the following relationship

$$\sigma_{12} = \frac{n_S z_S \Delta E}{N z_1},$$

where n_S is the number of atoms per unit surface at the interphase boundary; z_S is the number of bonds of the A-B type per atom; ΔE is the energy of solution; N is the Avogadro number; and z_1 is the coordination number.

The different expressions of the surface tension are given in [52-54], where the value of the surface tension is related to the heat of evaporation.

Theoretical investigations have shown that for liquids the surface tension is equal to the free energy per unit surface. For solids this is true only at high temperature, where the mobility of atoms is high and each change in the surface is accompanied by the establishment of an equilibrium. Also, surface tension in crystals is different in different directions, and this is a fundamental difference between crystalline bodies and liquids. The equilibrium shape of crystals is usually determined by the Gibbs-Curie-Wolff law, which determines the configuration of the crystal on the basis of minimum free surface energy

$$F = \sigma_1 S_1 + \sigma_2 S_2 + \cdots = \min.$$

This equation shows that in the equilibrium state single crystals must be limited by planes with a maximum reticular density. In most cases the configuration is more complex and the crystals acquire complicated shapes at the boundary with another phase.

Calculations show that when the crystal is very small a circular boundary is the most advantageous [80], and apparently for very small crystals with a size approaching the size of colloidal particles a sphere is the equilibrium shape. A round shape is of particular importance when the system is far from the equilibrium state. At the boundary of separation with the solid phase small crystals may have the shape of discs, ellipsoids, etc.

At the present time, a great deal of work is being done on the experimental determination of surface tension [81]. This problem will be examined in the following chapters.

CHAPTER 3

PHASE EQUILIBRIA AND PHASE DIAGRAMS

1. Equilibrium Conditions of a System

A system is in equilibrium when it is not acted upon by any external force and the parameters determining the state of the system remain constant over a long period of time. From the standpoint of the molecular-kinetic (statistical) theory, this means that there are no microscopic heat flows or particle flows in the system. In the equilibrium state the particles composing the system can move from one part of the system to another, although motion in any direction should be equally probable. When the temperature fluctuations are great the motion of a very large number of particles must be equally probable in all parts of the system. These two conditions are not equivalent, since the mechanisms of heat transfer and particle transport are different.

The first equilibrium condition — the absence of heat flow — is satisfied when the temperature is the same in all parts of the system.

The second equilibrium condition — the absence of particle flow — is satisfied if particles of the same kind have the same average free energy in all parts of the system.

In effect, the flow of particles is related to the work of transport, and consequently to the transformation of heat into work. The amount of free energy per atom of any component in an alloy with a given composition is determined by the magnitude of the chemical potential μ_i.

A flow of particles results from an energy difference in the direction of motion of the particles. Therefore, the absence of flow (and consequently equilibrium) occurs under the condition that

$$d\mu_i = 0.$$

This condition is sufficient for a multicomponent system with any number of phases. When this condition is satisfied the following conditions are automatically satisfied also [55]: that the following derivatives be equal:

$$\frac{\partial F_1}{\partial C_1} = \frac{\partial F_2}{\partial C_2} \cdots \tag{3.1}$$

that the curves representing the free energy of the coexisting phases have a common tangent:

$$F_1 - \frac{\partial F_1}{\partial C_1} C_1 = F_2 - \frac{\partial F_2}{\partial C_2} C_2 \cdots \tag{3.2}$$

These conditions are the consequence of the fact that the free energy of the system as a whole tends toward the minimum.

The considerations presented here can be illustrated by geometric analysis. Figure 13 shows curves of the free energy of the two phases $F_1(C)$ and $F_2(C)$ composing a given system.

In the equilibrium state the concentrations of the coexisting phases are C_1 and C_2. At these points the common tangent ab touches both curves, and consequently conditions (3.1) and (3.2) are satisfied. The change in the concentration of one of the phases leads to the

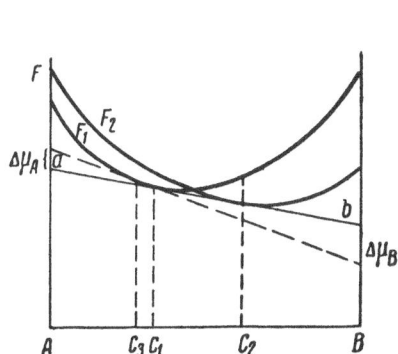

Fig. 13. Phase equilibrium conditions.

19

condition $\Delta\mu_A < \Delta\mu_B > 0$. As the result, the B atoms gain excess energy and move into areas of the system where the energy per atom of this kind is lower.

The A atoms move in the same way. As the result, the concentrations of the phases change so that the condition $\Delta\mu_A = \Delta\mu_B = 0$ is again satisfied. Thus, if the dependences of the free energy on the concentration and on the temperature for different states of the alloy are known, one can determine the composition of phases which are in equilibrium.

2. Effect of Intermolecular Interaction on the Shape of the Phase Diagram

The fundamental relationships of thermodynamics can be used for the construction of theoretical phase diagrams of different types only if the expressions of energy and the entropy of mixing are known. The first attempts in this direction were made in [56-61]. In what follows we give a brief description of the method of constructing phase diagrams which is used in [55].

A phase diagram is a geometric locus of points of phase equilibria in a system of temperature-concentration coordinates.

If the dependence of the free energy of the alloy in the liquid and solid states on the temperature and concentration are known, one can obtain an analytical expression which relates the temperature and the concentration of liquid and solid phases in equilibrium

$$F_l = C_1(1-C_1)V' + (1-C_1)V'_A + C_1 V'_B + kT\,[(1-C_1)\ln(1-C_1)$$
$$+ C_1 \ln C_1] - \frac{T}{N}[(1-C_1)S'_A + C_1 S'_B];$$
$$F_{TB} = C_2(1-C_2)V'' + (1-C_2)V''_A + C_2 V''_B$$
$$+ kT\,[(1-C_2)\ln(1-C_2) + C_2 \ln C_2] - \frac{T}{N}[(1-C_2)S''_A + C_2 S''_B].$$

Using the equilibrium conditions (3.1) and (3.2), we obtain

$$= \frac{C_1^2 V' - C_2^2 V'' + kq_A T_A}{S_A - \ln\dfrac{1-C_1}{1-C_2}} = \frac{(1-C_1)^2 V' - (1-C_2)^2 V'' + kq_B T_B}{S_B - \ln\dfrac{C_1}{C_2}}, \qquad (3.3)$$

where

$$S_A = \frac{Q_A}{kNT_A},$$
$$S_B = \frac{Q_B}{kNT_A}.$$

Here, Q and T are the heats of fusion and the melting points of the components.

Unfortunately, the equilibrium concentrations can be calculated by this equation only if the mixing energies of the components in the solid and liquid phases are known.

Therefore, at present these relationships are used to make the diagram more precise and to determine the exact position of the equilibrium line in the case where only small numbers of critical points are known. In connection with this, the use of computers is absolutely essential* for the calculation of tables or nomograms which make it possible to determine rapidly the value of the mixing energy for any alloy.

Let us now consider which of the properties of the components are responsible for the type of phase diagram.

---
* The calculation of phase diagrams for multicomponent systems is described in [64].

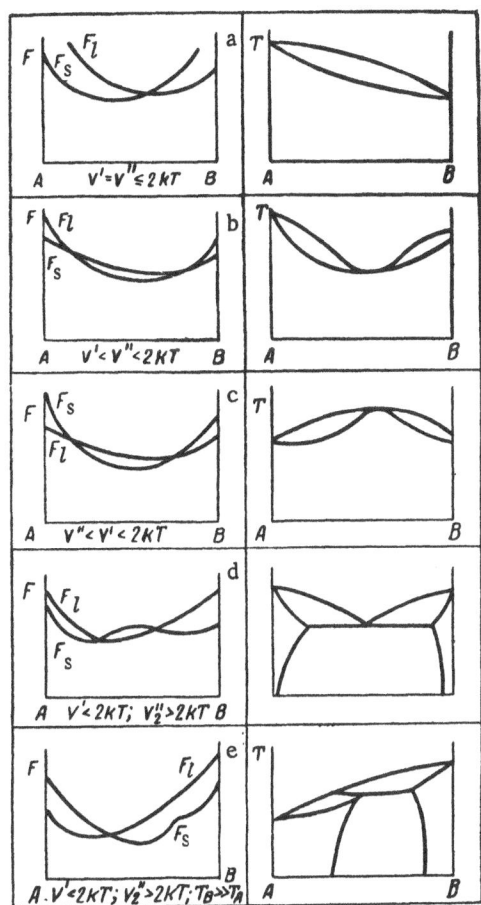

Fig. 14. Change in the shape of the phase diagram as a function of the values of V' and V".

From Eq. (3.3) one can deduce that a phase diagram is essentially determined by the value of the mixing energy and by the ratio of the mixing energy of the alloy in the solid and liquid states.

1. If $V' = V'' \leq 2kT$, the curves representing the free energy of the liquid and the solid phases have the form shown in Fig. 13. The distance between the curves changes with the temperature, and the points of tangency C_1 and C_2 of the common tangent ab indicate the compositions of the solid and the liquid phases in equilibrium.

In this case the phase equilibrium is described by the well-known cigar-shaped diagram. The mutual solubility of the components of the alloy is infinite in the solid and the liquid states (Fig. 14a).

2. $V' < V'' < 2kT$. The curve of the free energy is represented by the phase diagram in Fig. 14b.

3. $V'' < V' < 2kT$. The curves of free energy have the same shape as in the preceding case, except that the curvature of $F_{liq}(CT)$ is less than that of $F_{sol}(CT)$.

The $F(CT)$ curve and the phase diagram for this case are shown in Fig. 14c.

Within the temperature range investigated, the components of the alloy mix in unlimited amounts.

4. $V' < 2kT$; $V'' > 2kT$ when the difference between the melting points of the components is small. The curves of free energy and the phase diagram corresponding to this ratio of the components are shown in Fig. 14d.

The figure shows that the mixing of the components in the liquid state is infinite, while the solubility in the solid state is limited.

5. $V' < 2kT$; $V'' > 2kT$ when $T_B \gg T_A$. When the components of the alloy have these characteristics the phase equilibrium is described by a peritectic phase diagram and the $F_{liq}(CT)$ and $F_{sol}(CT)$ curves have the shapes shown in Fig. 14e.

We shall limit ourselves to these more frequent types of phase diagrams. We shall give some results of calculations of the mixing energies for a number of alloys for which phase diagrams are known in great detail.

Bi—Sb systems: a phase diagram indicating unlimited solubility in the solid and liquid states:

$V'_A = 46700$ cal/mole	$S'_A = 18.38$	$Q_A = 2600$ cal/mole
$V'_B = 41100$ cal/mole	$S'_B = 15.76$	$Q_B = 4750$ cal/mole
$V''_A = 44100$ cal/mole	$S''_A = 13.6$	$V' = 245$ cal/mole
$V''_B = 36350$ cal/mole	$S''_B = 10.5$	$V'' = 1384$ cal/mole

Ag—Cu system: a phase diagram of a eutectic type with limited solubility in the solid state:

$V'_A = 60700$ cal/mole	$S'_A = 10.20$	$Q_A = 2696$ cal/mole
$V'_B = 72800$ cal/mole	$S'_B = 7.97$	$Q_B = 3100$ cal/mole .

$$V_A'' = 58010 \text{ cal/mole} \qquad S_A'' = 8.02 \qquad V' = 3193 \text{ cal/mole}$$

$$V_B'' = 69700 \text{ cal/mole} \qquad S_B'' = 5.64 \qquad V'' = 5693 \text{ cal/mole}$$

The curves of free energy for this system are shown in Fig. 15.

Al–Si system: V' = (–5000)–(–8000) cal/mole (depending on the concentration) and V" = 5000–8000 cal/mole.

3. Solubility of the Phases of the Alloy

The amount of a component which dissolves in a given phase is determined by a complex interaction of such factors as kinetic energy and the energy of interaction between atoms of the components in adjacent phases.

Most of the components of alloys in the liquid state have unlimited solubility; the solubility is limited only when $V_{liq} > 2kT$.

In the solid state the miscibility of alloys is limited if $V_{sol} > 2kT$. The solubility limit of solid solutions in equilibrium is found by the position of the minimum on the curve describing the dependence of the free energy at different temperatures on concentration. The concentrations at the points of tangency of the common tangent ab (Fig. 11b) indicate the solubility limit of solid solutions α and β at the temperature T_1.

The region of separation can be obtained by using the spinodal curve, which is the geometric locus of inflection points of the $F_{sol}(CT)$ curve [61]. At the inflection point the second derivative of this function is equal to zero. Let us take the second derivative

$$\frac{\partial^2 F}{\partial C^2} = -2V'' + kT\left[\frac{1}{C} + \frac{1}{1-C}\right] = -2V'' + \frac{kT}{C - C^2} = 0,$$

and therefore

$$2C^2V'' - 2CV'' + kT = 0; \quad C_{12} = \frac{1}{2} \pm \sqrt{\frac{1}{4} - \frac{kT}{2V''}}.$$

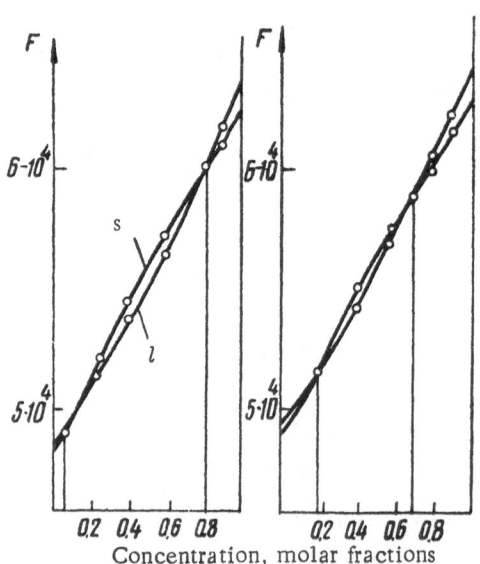

Using the value of V for the Ag–Cu system, one can mark on the phase diagram the theoretical region of the existence of α and β solid solutions (dashed line in Fig. 16).

The figure shows that the theoretical solubility of the components is somewhat too high.

It has been noted [62, 63] that during the formation of the solid solution the potential energy of the interaction of atoms is different from the mixing energy V because the deformation of the crystal lattice is not taken into account in calculating the mixing energy, and yet the crystal lattice is always deformed when atoms with different radii are mixed.

In the simplest case – when the elastic constants of the components are equal – one can take into account this singularity by representing the mixing energy V" as the sum of the energy of mixing V_0 and the elastic energy ε, which is always positive

$$V'' = V_0 + \varepsilon.$$

The change in the mixing energy during the formation of solid solutions was studied most extensively in [49]. That author gives the following expression for the mixing energy of components with the same lattice

Fig. 15. Variation of the free energies of the liquid and solid phases of Ag–Cu with the concentration [191]. a) At 1163°C; b) at 1093°C.

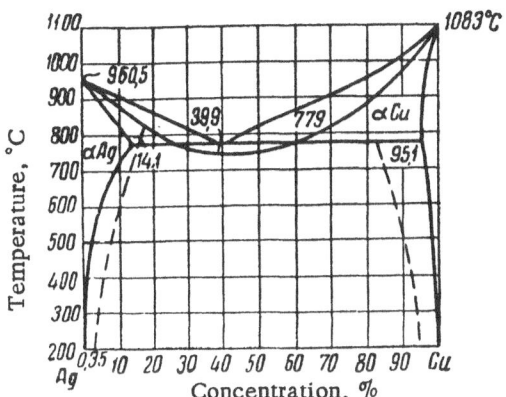

Fig. 16. Ag–Cu phase diagram.

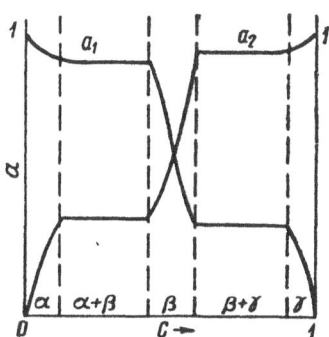

Fig. 17. Activities a_1 and a_2 in a system containing an intermediate phase [65]. α) Primary solution; γ) intermediate phase; β) primary solution.

$$V'' = V_0 + BK_A, \qquad (3.4)$$

where B is a constant and K_A is the modulus of hydrostatic compression.

Equation (3.4) shows that the solubility increases with increasing temperature because the value of K_A increases.*

Thus, the solubility limit of solid solutions is determined by the conditions of phase equilibrium. This is the same dynamic equilibrium in which atoms of different kinds can exchange at the phase boundary, and the number of atoms passing into one phase is equal to the number of atoms of the same kind returning to the initial phase.

The equations given above do not take into account the size of the structural components.

If the system consists of a mother solution in which the second phase is distributed in the form of small round crystallites, then dynamic equilibrium is established at the boundary between the mother solution and the dispersed phase. This equilibrium is somewhat different from the equilibrium at a plane boundary. On the convex surface of the dispersed phase the atoms are retained by a smaller number of neighbors, and are therefore easier to detach, and can pass into the surrounding solution. Therefore, the solubility of the primary solution depends to a great extent on the curvature of the boundary separating the adjacent phases. This peculiarity can be taken into account by using the Thompson formula, which is derived rather strictly on the basis of the molecular-kinetic concept

$$C = C_{\sim} e^{\frac{2\sigma MV}{RTr}}. \qquad (3.5)$$

Here C is the concentration of the mother solution adjacent to the particles of the dispersed phase with a radius r; $C_{\sim}$ is the equilibrium concentration in the case of a plane boundary; σ is the surface tension at the phase boundary; M is the molecular weight of the dispersed phase; V is its specific volume; R is the universal gas constant; and T is absolute temperature.

4. Intermediate Phases on the Equilibrium Diagram

The number of phases in equilibrium at a constant pressure in an alloy consisting of several components is determined by the Gibbs phase rule

$$f = k + 2 - n.$$

---

* For most metals $K_A = K_0 + B_1(T/T_S)$ (where B_1 is a constant and T_S is the melting point), and consequently,

$$V \cong V_{0_1} + B_{0_1} \frac{T}{T_S}.$$

23

In this relationship f is the number of thermodynamic degrees of freedom or the number of independent variables which can be arbitrarily changed without disrupting the equilibrium of a heterogeneous system. For alloys, these independent variables are usually temperature and concentration; k is the number of components in the alloy; n is the possible number of phases in equilibrium.

If we do not take the vapor phase into account, then for a unicomponent system the maximum possible number of phases in equilibrium is equal to two.* This is either the equilibrium between the liquid and the solid phase or an equilibrium between two different modifications of the components in the solid phase. In either case, the system is invariant, i.e., it can be in equilibrium only at constant temperature equal to the melting temperature or to the temperature of polymorphic transformation.

In binary systems the maximum possible number of phases in equilibrium is equal to three, and in this case the equilibrium exists either at a constant eutectic, peritectic, or eutectoid temperature and the concentrations of phases in equilibrium must be quite definite.

However, intermediate phases can also form in binary systems. From the viewpoint of the thermodynamics of equilibrium states, and to satisfy the phase rule of the phase diagrams of such systems, the system must be divided into several systems in which the intermediate phase plays the role of independent component. However, a system consisting of two components and one intermediate phase cannot be regarded as a three-component system.

In fact, according to the phase rule, four phases can exist simultaneously in a three-component system in equilibrium. In systems such as carbon and iron, for example, these phases are the solid solution based on iron (austenite), the liquid solution, iron carbide, and graphite. However, these four phases cannot be in equilibrium simultaneously, and the phase rule in this case is inapplicable.

Let us investigate such a case by using geometric analysis [65]. Let us assume that the intermediate phase of concentration C_1 can form in a two-component alloy. Usually, the intermediate phases exist only within narrow ranges of concentration, since they are highly ordered. Therefore, the dependence of the activity of components A and B on the concentration is shown by the curves in Fig. 17. In the region of heterogeneous equilibrium $\alpha + \beta$ and $\beta + \gamma$ the activities of atoms of the same component must be the same in both phases, since the condition $\Delta\mu_{1-2} = RT \ln a_1/a_2 = 0$ must be satisfied; and consequently $a_1 = a_2$.

The figure shows that this condition cannot be satisfied in the case of equilibrium of both primary solutions and the intermediate phase because the activity increases with increasing concentrations of any solutions (not ordered) which include the primary solid solutions and the ordered intermediate phases. In the heterogeneous region the activity does not change because the concentrations of phases remain constant.

The free energy F of a heterogeneous system consisting of several phases with free energies F_1, F_2, F_3, etc. respectively, is linearly dependent on the concentration, since $F = C_1F_1 + C_2F_2$. For a binary alloy, $F = CF_1 + (1-C)F_2$, where C is the concentration of component 1 in the alloy. Graphically, the change in the free energy of the binary alloy containing intermediate phases is shown by the curve in Fig. 18 in the case where the mutual solubility of the components is low. The different straight segments correspond to the values of the free energies of the mixtures of phases. These values of free energy are minimum for the alloy with a given composition.

If the system forms a series of solutions and intermediate phases, one usually uses a hypothetical curve of the free energy of each phase (Fig. 19). In this figure the $F_1(C)$ curve corresponds to the melt, $F_2(C)$ to the α-solid solution, $F_3(C)$ to the intermediate phase γ, and $F_4(C)$ to the primary solid solution β. The common tangents to these curves make it possible to determine the number and the composition of co-existing phases. The minimum value of the free energy is the main criterion for determining the highest degree of equilibrium of the system of phases for a given composition of the alloy.

5. The Approximation of the Theory

It is extremely difficult to draw phase diagrams from experimental data. The principal difficulty is the precise determination of the liquidus and solidus lines for alloys with high melting points. Therefore, many attempts have been made to calculate these diagrams [66-70, 51].

---

* If we take the pressure into account then three phases – liquid, solid, and gaseous – can exist in equilibrium at the triple point of the unicomponent system.

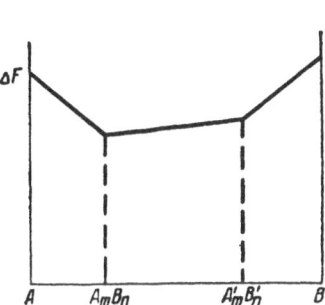

Fig. 18. Variation of the free energy of a system containing chemical compounds.

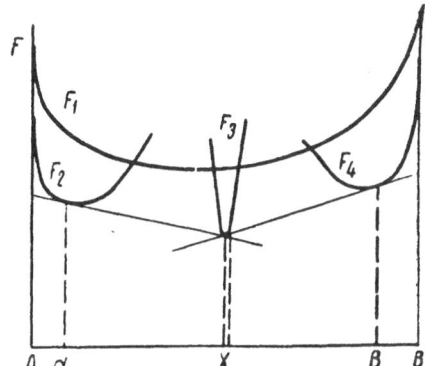

Fig. 19. Variation of the free energy of a system with an intermediate phase.

Different types of phase diagrams were described in [61]. The application of the theory described above to the calculation of relatively simple phase diagrams gives results in good agreement with the experimental results. In many cases these calculations make it possible to check and refine the data on phase equilibrium in different systems [71]. However, in more complex systems containing intermediate phases and also in the case of transformations in solids the application of the theory is difficult.

Essentially, the difficulty is that the expression of the mixing energy is approximate and there are no reliable data on the energy and entropy of mixing of component and intermediate phases. Work is being done to find more precise expressions of these magnitudes and also to determine the relationship between them and the known characteristics of the system, such as elastic modulus, surface tension, etc.

Naturally, the increasing number of corrections makes calculation of the phase diagram even more complicated. Even in the simplest cases the calculations are extremely long. In this respect, computers are very promising, since we cannot expect any simplification of the theory of such a complex process as the formation of the structure of an alloy.

The attempt to find simpler calculations which can be applied to a large number of phase diagrams necessarily leads to semiempirical solutions of this problem. Thus, in [72] the author uses the model of a "subregular" solution where he formally introduces the dependence of the free energy on the concentration, which extends the possibilities of calculating phase diagrams.

In [73] the mixing energy is expressed in terms of thermodynamic activity. The mixing energy and entropy are calculated by using experimental data on the heat content of alloys. This method makes it possible to calculate phase diagrams of systems forming intermediate phases. Thus, the theory of constructing a phase diagram has been formulated quite recently and the conclusions are only a first approximation of the real equilibrium conditions in an alloy. The theory is a success, but further development of the theory is one of the current problems of the physics of metals.

CHAPTER 4

NONEQUILIBRIUM STATE

1. Metastability

The concepts of "metastable" and "labile" solutions were introduced as early as the beginning of this century. It is well known that the solubility of any substance in a mother solution depends in most cases on the temperature. With decreasing temperature the solution becomes first supersaturated and then the excess phase begins to precipitate. The different states of the solution are represented schematically in Fig. 34 (p. 63). The BC line represents the solubility limit. Below this limit the solution is supersaturated; below the B_1C_1 line (metastability limit) the dissolved substance begins to precipitate. Here (between lines BC and B_1C_1) the solution will be in the supersaturated state and it is called metastable.

Below line B_1C_1 the solution is in unstable equilibrium, which is called labile.

From the thermodynamic viewpoint these concepts have a much wider meaning. If some system is transformed from one state into another, complete equilibrium can be reached only if the rate of change of the parameters determining the state of the system is infinitely slow. In this case the final state has a minimum free energy, it is called stable, and the transformation occurs under equilibrium conditions.

However, most changes in the structure of the alloy, be it hardening or heat treatment, are accompanied by changes in the principal parameters during a finite (although often very short) period of time. As the result, the transformation from one state into another occurs under nonequilibrium conditions, and this may lead to the creation of intermediate states with high free energies. At low temperatures, when the mobility of atoms is low the system can remain for a very long time in such a state until some external action forces it into the stable state.

The system having a free energy higher than the energy which it would have in the equilibrium state at the same temperature and composition is called the metastable state. Thus, the measure of the metastable state is the excess free energy as compared to the energy of the system in the equilibrium state. The time in which the system passes from one state to another is called the relaxation time. During this time the system is in the labile state.

These concepts cover all the possible intermediate states in alloys that can exist at different cooling rates.

Let us consider the different types of metastable states existing in alloys.

1. An increase in the dispersity of the crystallites in the alloy and the branching of the crystals lead to an increase in the total surface of separation between phases. This increases the free energy of the alloy as compared to single crystal alloys or alloys in which the crystallites have a more regular shape. The most favorable shape is determined by the minimum value of the surface tension at the separation boundary and by the minimum value of the surface of separation. The metastability can be measured by the excess of free energy due to the change of these values.

2. The second type of metastability results from the occurrence of supersaturated solutions. These can be formed in the liquid or solid state. In the latter case there is also deformation of the crystal lattice. In both cases the total free energy of the system increases, and this increase is due to the increase in the chemical potentials of all the atoms of the alloy.

3. There are quite often cases when the number of phases in the alloy do not obey Gibbs' phase rule. In cast iron, for example, austenite, cementite, and graphite can coexist simultaneously. Under equilibrium conditions (at the same temperature) this system consists only of austenite and graphite.

In this case the metastable system has an excess free energy because the cementite itself has a large free energy as compared to that of graphite. Thus, the increase in the free energy of the whole system in this case is due to the presence of a metastable phase which has an excess free energy.

These cases exhaust the possible kinds of metastable alloys. They can be divided into two types. The first type of metastability is related to the change in the total surface energy of the system. It is determined by

$$\Delta F = \Delta \sigma \Delta S, \tag{4.1}$$

where $\Delta \sigma$ and ΔS are respectively the excess surface energy and the excess total length of the separation boundary as compared with those in the equilibrium state. The second type of metastability is related to the change of the free energy relative to the volume and can be calculated from the variation of the partial chemical potential of the atoms of the phases:

$$\Delta \mu_1 = RT \ln \frac{C_1}{C_0}, \tag{4.2}$$

where C_1 and C_0 are respectively the concentrations of the supersaturated and equilibrium solutions.

If supersaturation is accompanied by internal stress (for example, in solid solutions) one must add to this value the energy relative to elastic deformation; at the present time it is difficult to account for this value.

The total excess free energy of a two-phase system with all types of metastability is determined by

$$\Delta F = C_1 \Delta \mu_1 + C_2 \Delta \mu_2 + \sigma \Delta S + \varepsilon, \tag{4.3}$$

where ε is the energy relative to the internal stress; C_1 and C_2 are the concentrations of the components of the alloy.

In a system consisting of a mother solid solution and a dispersed phase with globular crystallites the energy increases not only because of the increase of the length of the boundary but also because of the increase in the solubility of the surrounding solution.

The solubility of the mother phase is determined by the Thompson equation. If we substitute the value $C/C_\sim$ from Eq. (3.5) into (4.2) we obtain the following relationship

$$\Delta \mu_B = \frac{2\sigma MV}{r}, \qquad \Delta \mu_A = \frac{-2\sigma MV}{r}$$

Here the minus sign indicates that the A atoms are on the convex boundary (−r). The change in the free energy related to the change in the solubility (Fig. 20) is

$$\Delta F = C\Delta \mu_A + (1-C) \Delta \mu_B = \frac{2\sigma MV}{r} (1-2c)].$$

According to Eq. (4.3), if σ is constant and $\varepsilon = 0$, we have

$$\Delta F = \sigma \left[\Delta S + \frac{2MV}{r} (1-2c) \right].$$

The second term in the square brackets is important only when r is very small and the solubility of the components is very low.

Using a construction similar to that shown in Fig. 19, one can show that the presence of the metastable state leads to the following change in the free energy of the system

$$\Delta F = (1-C) \Delta \mu_B - C\Delta \mu_A + \Delta F\beta,$$

where ΔF_β is the difference between the free energies of the stable and metastable phases, which can also be expressed in terms of the difference in partial chemical potentials.

If we take into account the fact that the surface tension is also related to the jump in chemical potential at the separation boundary, then we may assume that the metastability of the system is determined by the total excess chemical potential of the particles of the system.

27

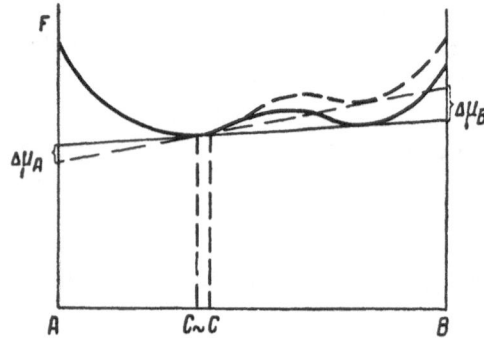

Fig. 20. Variation of the solubility with increasing dispersity of the crystals of the second phase Fe(C) for a system with a dispersed phase.

2. Some Information on the Thermodynamics of Irreversible Processes

Under certain conditions irreversible processes can transform an alloy with a given degree of metastability into the stable state.

Processes are irreversible when they proceed spontaneously in only one direction. Without the participation of external forces (e.g., deformation) they can only decrease the degree of metastability.

Among irreversible processes there are: 1) the uniformization of the concentration within the crystallites; 2) the decrease of the length of the separation boundary as the result of recrystallization, coalescence, or spheroidization; 3) decrease of the concentrations of solid solutions in the process of crystallization and recrystallization; 4) dissolution of metastable states as the result of the increase of stable states (in graphitization of white cast iron, for example); 5) relief of internal stress; 6) decrease of the surface tension during polyhedrization.

In all these processes transport is the principal phenomenon: diffusion in all its aspects, heat conductivity, internal friction, etc.

At first glance it seems that these phenomena have nothing in common with the thermodynamics of equilibrium processes. In reality, a detailed study of these phenomena shows that they are based on the principles of the molecular-kinetic theory of the structure of matter.

Since the formation of the structure of alloys is accompanied by diffusion and heat conductivity, we shall discuss these phenomena in great detail.

The application of classical thermodynamics to irreversible processes is based on the concept of local equilibrium. Any homogeneous body can be separated into small parts containing a sufficiently great number of particles that these parts of the system have the same temperature and the same chemical potential as the whole system.

The establishment of equilibrium in the whole system begins from some local equilibrium, i.e., equilibrium in small parts of the system. For instance, the uniformization of the temperature or concentration in the whole system is much slower than in microvolumes. The enormous difference in the rates of establishment of total and local equilibrium allows us to speak of a definite temperature and chemical potential of atoms in small parts of the system even in the nonequilibrium state. Therefore, although not all the system is in equilibrium we may assume that in small volumes (containing, however, a sufficiently large number of particles) there is local equilibrium and there is a microscopic change in the parameters between one small system of particles and another.

In the case of the nonequilibrium state it is very important to know how the system changes from one area to another or, roughly speaking, the "slope" of the change in the nonequilibrium state. This value is represented by the gradient of chemical potential *

$$\nabla \mu_i = \frac{\partial \mu_i}{\partial x}, \qquad (4.4)$$

i.e., the change of chemical potential at the distance x. The physical meaning of this value is reduced to the fact that it determines the force acting on the particles in a given area of the system [74]. This force is in the direction from the area of high chemical potential to the area with a low chemical potential. In the simplest case of a linear dependence we have

---

* We assume that $\partial T/\partial x = 0$.

$$\overline{F} = -\frac{\Delta \mu_i}{\Delta x},$$

where $\overline{F}$ is the average force acting on the atoms of the alloy in a region in which there is a difference in temperature, concentration, stress, etc.

If this force is sufficient to overcome the binding forces between atoms resulting from atomic interaction this force causes the flow of atoms in the direction of areas with lower energies.

Thus, one can use thermodynamics to describe the causes of the displacement of atoms in the direction of areas with lower energies.

3. Problems of Kinetics

Irreversible phenomena which occur in an alloy as the result of some force are characterized by flows (thermal or diffusional, for example).

These values usually represent the rate of change of the corresponding parameters determining the state of the system. Let us first investigate the general laws relative to the direction of the displacement of atoms in the alloy, i.e., the process of diffusion.

The flow of particles is defined as the product of the number of particles per unit volume n and the average rate of motion:

$$q = nv.$$

This magnitude can be expressed in terms of the average force F acting on the particles; $q = nbF$, where b is the mobility of particles. This magnitude is equal to the rate of motion of particles when the force acting upon them is equal to unity. If we substitute F from (4.4) into this expression, we obtain

$$q = nb\nabla\mu = -nb\frac{\partial \mu}{\partial x}$$

However, to calculate diffusional flow one uses not the gradient of chemical potential but the concentration gradient of $\partial n/\partial x = \nabla n$, which is much easier to determine experimentally.

In this case the flow will be $q = -D\partial n/\partial x$. The proportionality coefficient D is called the diffusion coefficient and represents the flow of particles when the concentration gradient is equal to unity.

Using both equations, one can determine the relationship between the diffusion coefficient and the chemical potential

$$D = nb\frac{\partial \mu}{\partial n}. \qquad (4.5)$$

In most cases it is necessary to know not the flow but the amount of the substance Q diffused during a certain period of time

$$Q = \int_0^t q\,dt.$$

The solution of this problem is more or less complex, depending on the conditions under which diffusion occurs. The most important of these conditions are the configuration of the surface through which diffusion occurs and the behavior of the concentration gradient during the process.

If the concentration gradient does not change in time then such a flow is called steady; if the gradient does change, the process is called nonsteady and the change in the concentration at a given point of the system is determined by the relationship $\partial C/\partial t = D\,\partial^2 C/\partial x^2$, while in the steady state it is

$$D\frac{\partial^2 C}{\partial x^2} = 0. \qquad (4.6)$$

The solution of this equation for different cases of diffusion is given in the literature [75, 76]. The equations for heat flow w have the same form

$$w = -\chi \nabla T.$$

Here χ is the coefficient of heat conductivity; $\nabla T = \partial T / \partial x$ is the temperature gradient. For nonsteady heat flow the change in the temperature at a given point in the system is expressed by

$$\frac{\partial T}{\partial t} = \frac{\varkappa}{c_V} \frac{\partial^2 T}{\partial x^2},$$

where c_V is the heat capacity at constant volume and χ / c_V is called the coefficient of thermal conductivity.

4. Coefficients of Diffusion and Heat Conductivity

Equation (4.5) shows that the diffusion coefficient depends on the concentration and temperature.

Many attempts have been made to calculate the dependence of the diffusion coefficient on other constants characterizing the state of a system [11, 77-79]. These calculations show that the diffusion coefficient depends also on the mechanism by which the atoms pass from one position to another.

There are several basic mechanisms of the displacement of atoms: a) simple exchange of places between neighboring atoms; b) circular displacement of a chain of atoms; c) movement through successive intersites; d) movement through vacancies. In many cases these mechanisms occur simultaneously.

The dependence of the diffusion coefficient on the temperature is well represented by the empirical formula

$$D = D_0 e^{-\frac{Q}{RT}},$$

where D_0 is the preexponential term; Q is the activation energy or the energy of the loosening of the lattice.

The various theoretical calculations are reduced to the expressing D or Q in terms of known constants. So far, these calculations have been successfully applied only to a rather small number of specific cases of diffusion.

The following empirical law is well known: The less compact the lattice and the more defects it has, the greater the diffusion coefficient.

From this rule one can derive a number of specific empirical results, among which is the following: the diffusion coefficient increases with the concentration of the alloy. For example, the diffusion of carbon in γ-iron is given by

$$D = (0.04 + 0.08C) e^{-\frac{31350}{RT}}$$

Very often, the diffusion rate is greater in the surface layer and along the grain boundary than in the crystallites themselves. Apparently, this is never true in the case of formation of penetration solid solutions – for example, in alloys of iron with small amounts of carbon nitrogen, or hydrogen.

A third component has a great effect on diffusion [81, 82]. The diffusion of carbon in austenite alloyed with elements that do not form stable carbides (Co, Ni) decreases Q but affects D_0 very little, while elements such as Mn, Cr, W, and Mo increase the heat of loosening (in that order) and also increase the value of D_0 [83].

Sometimes, the mutual diffusion of two metals is accompanied by intense formation of vacancies in pairs [84, 85].

Different types of local deformations of the lattice play an important role in diffusion processes [86]. At high temperatures the mobility of atoms is high and these defects usually have little effect. At low temperatures, where ordinary displacement of atoms is slowed down, atoms move only where there are deformations.

Deformations are usually concentrated around phase boundaries because the coefficients of volume expansion of different phases are different. Also, they can be related to the complex distribution of stresses, and consequently to local concentrations. Areas with local stresses may arrange themselves in a certain direction and the diffusional flow will follow this direction.

Phase transformations occurring during solidification of a liquid are also accompanied by diffusional separation of components.

It is difficult to study diffusion in liquids because it is accompanied by convectional mixing of atoms, which often occurs at high rates. The pure diffusion process follows the same laws as in the solid state.

The diffusion rate of atoms in the liquid state must be considerably higher because liquids have many defects and because diffusion occurs at a higher temperature.

The diffusion coefficients of different elements in liquid iron were measured by various investigators [87–89]. In iron containing 0.03–3.5% Mn, Si, P, S, Co, and C the diffusion coefficient decreases from approximately $7.9 \cdot 10^{-5}$ to $6.7 \cdot 10^{-5}$ cm^2/sec.

The conditions under which diffusion occurs during phase transformation are very complicated. Aside from the complex configuration of the separation boundary and the change of its position, the heterogeneous fields of stresses and temperatures make diffusion very complex. The heterogeneous stress field occurs because of the difference in the specific volumes of the growing phase and the solid solution surrounding it. The evolved heat of transformation, which directly affects the distribution of the temperature, the concentration, and the diffusion coefficient itself, plays an even greater role.

At present, methods have been worked out which make it possible to calculate diffusional processes in which some of the secondary phenomena are taken into account [82].

A body is also in a nonequilibrium state if its temperature is different from point to point. The energy flow resulting from the uniformization of temperature is proportional to the temperature gradient. For a steady flow one obtains an equation similar to Eq. (4.6):

$$\varkappa \frac{\partial^2 T}{\partial x^2} = 0.$$

The coefficient of heat conductivity determines the amount of heat which passes through a surface 1 cm^2 in the case of a temperature gradient of 1° at a distance of 1 cm.

The heat conductivity characterizes the rate of change in the temperature.

The heat conductivity coefficient depends on the temperature. The general form of the variation of heat conductivity with temperature is:

$$\varkappa (T) = \varkappa (1 + \alpha t),$$

where α is the temperature coefficient of the heat conductivity.

The diffusion rate and the rate of heat transfer are very different from each other, since a considerable part of the heat is transferred by the electron gas. A constant temperature must be maintained in all parts of the solvent to study diffusion under conditions where no phase transformations occur. The study of diffusional flows, particularly the distribution of concentrations in the solution during the process of phase transformation, is a much more complex problem. Diffusional transformations are usually accompanied by the evolution of heat at the transformation front and this, in turn, complicates the distribution of the temperature field directly at the front.

Furthermore, equilibrium conditions are established in small volumes very rapidly, and therefore the distribution of concentrations must correspond to the distribution of temperature directly at the boundary where the phase transformation occurs.

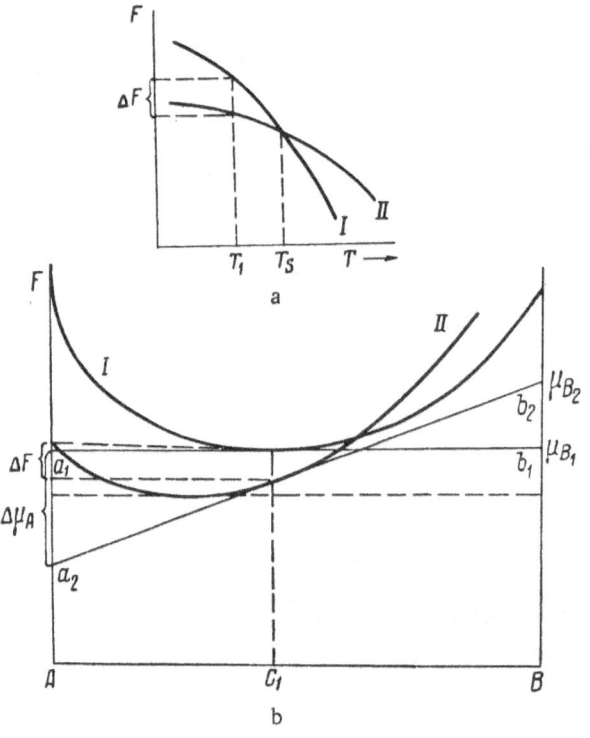

Fig. 21. Dependence of the free energy on: a) The temperature of a unicomponent system; b) the composition during supercooling.

5. Supercooled and Supersaturated Systems—Reaction Rates

The metastability of a given state of an alloy can be determined only with respect to the stable state. The system is metastable only by virtue of some properties related to the mechanism of the transformation into the stable state itself.

As a rule, a system passes from one state to another under the effect of a moving force, which directs the atoms of the system, creates flows of atoms, and determines the rate of transformation. This force is counteracted by the mobility of the atoms of the alloy, which is itself directly related to the temperature.

Only for a given ratio between these two factors can the system be in a metastable state for a long period of time.

If the forces which determine the transformations of systems from one state to another are small with respect to the forces which maintain the level of the random motion of atoms, then the system can remain in the metastable state even at high temperatures. If, however, the transformation forces are high and the mobility of the atoms is negligibly small, which occurs at low temperatures, then the system cannot pass into the stable state, since no significant flow of atoms (responsible for such a transformation) is possible.

Moving forces are gradients of partial chemical potentials in a thermodynamic system.

If we compare the effects of forces at a finite distance, their magnitudes can be determined by the difference in the partial chemical potentials in a given state

$$\overline{F} \sim \Delta\mu_i = kT \ln \frac{a_1}{a_2}.$$

Let us first examine the metastable state in a unicomponent system. In such a system the following metastable states can occur: supercooled liquid up to an amorphous solid state; high-temperature polymorphic modification stabilized at low temperature by quenching.

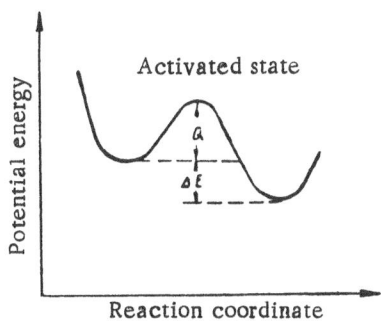

Fig. 22. Variation of the potential
energy during the reaction.

In a unicomponent system the partial chemical potential is equal to the free energy per atom of the system. Figure 21 shows the dependence of F on the temperature for any two states of the system (I, II). Let us assume that I is the liquid phase and II the solid phase. When the temperature of the liquid decreases to T_1 the liquid passes into the metastable state, in which the forces acting in the direction of the transformation into the solid state are determined by ΔF; the magnitude of ΔF increases as the supercooling of the system increases. For low degrees of supercooling one usually uses the approximate first-degree Taylor polynomial [11]:

$$\Delta F = \frac{L\Delta T}{T_S},$$

where L is the heat of transformation; ΔT is the degree of supercooling; and T_S is the melting point. For a better approximation one can use the second-degree polynomial

$$\Delta F = \frac{L\Delta T}{T_S} - \frac{c_2 - c_1}{2T_S}(\Delta T)^2, \qquad (4.7)$$

where $c_2 - c_1$ is the difference in the heat capacity in the liquid and solid states.

This equation shows that the external factor which determines the degree of metastability in a unicomponent system is supercooling. Supercooling means the difference between the equilibrium temperature of phases and the temperature of the metastable state.

Let us now examine a two-component system with unlimited solubility in the solid and liquid states. Figure 21b shows the dependence of the free energy per atom on the temperature and composition in the liquid and solid states. At some temperature T the liquid solution with a composition C_1 is in the metastable state, since $F_1(C_1) > F_2(C_1)$.

The transformation of the liquid state into the solid state occurs with a gain of free energy per atom of solid solution $\Delta F = F_1(C_1) - F_2(C_1)$; in terms of final states, such a direct transformation is advantageous and must occur at a high rate without diffusional redistribution. However, the analysis of the forces acting on the atoms leads to somewhat different conclusions.

In fact, the partial chemical potentials of atoms in a liquid solution are determined by the segments on the A and B axes cut by the tangent a_1b_1, and for the solid state of the same composition by the segments cut by tangent a_2b_2. Consequently, the A atoms pass into the solid solution with a lower energy, since they are acted upon by the forces $F_A \sim \Delta\mu A = \mu A_1 - \mu A_2 > 0$, while the B atoms are pushed out of the solid solution, since $F_B \sim \Delta\mu B = \mu B_1 - \mu B_2 < 0$.

If the solid solution with a composition C_1 is surrounded by a liquid with the same composition at temperature T, then only the A atoms will pass from the liquid to the solid phase, and thus the concentration of the solid phase will decrease until the liquid becomes supersaturated in atoms of the B component. This will occur when all the points of the tangent a_2b_2, turning around curve II in a clockwise direction (because the concentration of the solid phase decreases), are above line a_1b_1.

In a two-component system the metastable state is also related to supercooling, but in this case an important factor is supersaturation with one or both components. Although in this case the degree of metastability is determined by the difference between the free energy in the initial and final states, transformation into the stable state is much more complex than in a unicomponent system. The transformation depends to a great extent on the supersaturation of the system with one or the other component.

Any process in metals and alloys related to a change in free energy is usually the result of a large number of elementary movements of the atoms composing the system. The transformation of atoms from one state into another is shown in the diagram in Fig. 22. Here, Q is the activation energy necessary to overcome the potential barrier. Atoms possessing this energy, which are at the top of the barrier, are called activated. The difference between the potential energies in the final states ΔE is called the reaction heat.

The reaction rate is determined by the products of the number of activated atoms (activated complexes, transition states) per unit volume and the frequency of crossing of the potential barrier [290].

The number of activated atoms n is determined by $n = n_0 \, e^{-Q/RT}$, where n is the total number of atoms. The reaction rate is expressed by

$$K = n\nu e^{-\frac{Q}{RT}}.$$

In this expression the frequency of oscillation $\nu = kT/h$, where h is Planck's constant.

If we take into account the fact that the atoms have different potential energies before and after passing the potential barrier, then the expression of the reaction rate has the form

$$K = n\nu e^{-\frac{\Delta F}{kT}},$$

where ΔF represents the free energy of activation per atom.

The application of quantum mechanics and statistics to the determination of the preexponential term and the free energy of activation led to the creation of a theory which was called the theory of absolute reaction rates (theory of the transition state). The theory of absolute reaction rates is used to determine the rate of any process of regrouping of atoms in metals and alloys. The comparison of the rates of different processes makes it possible to determine the most probable mechanism of the transformation of a system from one state into another.

The application of the conclusions of this theory to the process of diffusion, which often determines the transformation rate in alloys, leads to the following expression of the diffusion coefficient: $D = \lambda^2 K$, where λ is the distance between two neighboring equilibrium positions of the diffusing atoms.

From this we can obtain an expression for the dependence of the diffusion coefficient on the thermodynamic activity (a) of atoms of the diffusing element in a given solution [291]:

$$D_A = D_A^* \frac{d \ln a_A}{d \ln C_A},$$

where D_A^* is the diffusion coefficient in an ideal solution; C_A is the molar fraction of the A component.

This expression is particularly valuable for the study of the kinetics of phase transformations because it accounts for the fact that the motive force of diffusion is not the concentration of the component but the thermodynamic activity of the atoms. This is confirmed by cases of equilibrium between liquid and solid solutions with different concentrations of the components.

In such a phase equilibrium diffusion from one phase into another is not the result of the equality of thermodynamic activities when the concentrations are very different, but depends to a great extent on the supersaturation of the system with one or the other component.

CHAPTER 5

MECHANISMS AND KINETICS
OF PHASE TRANSFORMATIONS

Theoretical investigations made by Gibbs as early as 1878 [269] showed that the occurrence of a new phase is determined by the work of formation of the interphase separation boundary. The crystallization of ingots from low-melting-point systems observed in [270] and investigated in [271] led to an explanation of crystallization as a process of nucleation and subsequent growth of the nuclei of the new phase after the system has reached a certain degree of supercooling. A few decades later, the development of statistical physics and the accumulation of a large number of experimental data led to further development of this theory, first in [272], then in [273, 274], etc.

At the present time, the principal results of this theory are treated in many textbooks on the physics of metals and metal science, and therefore we shall use only the final equations of the theory and emphasize their physical meaning.

1. Formation of Nuclei

Let us first examine the singularities of the mechanism of the formation of nuclei of a new phase during the transformation of the liquid state (I, Fig. 21) into the solid state (II, Fig. 21). According to the laws of thermodynamics, the change in the free energy of a system ΔF must be the sum of the changes in the free energies of atoms resulting from the transformation from one phase into another ($\varphi_2 - \varphi_1$) and the free energy at the interphase separation boundary:

$$\Delta F = V_1 \frac{\rho}{M} (\varphi_2 - \varphi_1) + k V_1^{2/3} \sigma. \tag{5.1}$$

Here V_1 is the volume of the solid phase and $k V_1^{2/3}$ is its surface. Since $\varphi_2 < \varphi_1$ is lower than the crystallization temperature, the free energy can decrease only when the absolute value of the first term in the equation is greater than that of the second term. Consequently, function (5.1) has a maximum when $V_1' = [2 k \sigma M / 3\rho(\varphi_1 - \varphi_2)]^3$. This is the so-called critical volume of the newly formed phase.

Therefore, the critical size of the nucleus necessary for spherical and cubic crystals is

$$r^* = \frac{2\sigma M}{(\varphi_1 - \varphi_2)\rho} .$$

In the case of a low degree of supercooling $\varphi_1 - \varphi_2 \approx L\Delta T/T_S$, and consequently

$$r^* = \frac{2\sigma M T_S}{L\rho\Delta T} . \tag{5.2}$$

The work of formation of a nucleus of such a shape is

$$A = \Delta F = \frac{1}{3} \sigma S^*,$$

where S^* is the surface of the critical nucleus.

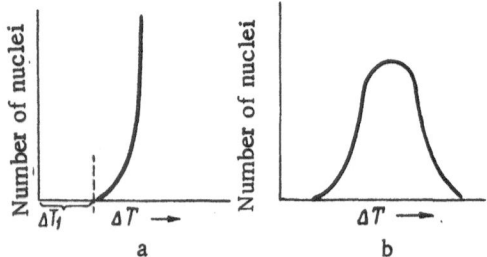

Fig. 23. Rate of formation of crystallization centers. a) In metal-like liquids; b) in viscous liquids.

The size of the critical nucleus in the case of the precipitation of a substance from a supersaturated solution is

$$r^* = \frac{2\sigma M V}{RT \ln \dfrac{C}{C_\sim}},\qquad (5.3)$$

where $C_\sim$ is the equilibrium concentration of the solution at the transformation temperature T; C is the concentration of the supersaturated solution.

It is assumed that the critical nuclei can be in equilibrium with the supercooled liquid. This means that the temperature and the concentration of the liquid at the boundary with the nucleus are the same as away from it. If because of some accidental addition of atoms the size of the nucleus increases, the equilibrium is shifted toward the growth of the nucleus; if the nucleus loses some atoms the nucleus begins to dissolve. Thermodynamic calculations do not describe the mechanism of the creation of the nucleus. On the basis of the molecular-kinetic theory, one can assume that the nucleus is created as the result of fluctuations. The probability of fluctuations is determined by the relationship

$$p = p_0 e^{-\frac{A}{kT}}.$$

The rate of formation of centers, taking into account the variation of the viscosity of the liquid with temperature, is

$$I = K e^{-\frac{A}{kT}} e^{-\frac{U}{kT}},\qquad (5.4)$$

where A is the work of formation of the nucleus and U is the activation energy.

Complex calculations are needed to find the value of the constant K. It is close to the number of molecules in the system.

The theory has been checked experimentally several times. The most general and thorough examination was made in [12].

It turned out that in the case of crystallization of metal alloys one can neglect the second exponent in (5.4). Then, the rate of formation of centers in a unicomponent supercooled melt is determined by the relationship

$$I = K_t e^{-\frac{B\sigma^3}{T\,(\Delta T)^2}},\qquad (5.5)$$

where

$$B = 32\left(\frac{M}{\rho}\right)^2 T_S^2\,\frac{1}{L^2}.$$

The variation of the rate of formation of centers with the supercooling temperature is shown in Fig. 23. The figure shows that the formation of crystallization centers requires a certain minimum supercooling $\Delta T_1 = T_0 - T_1$, which is called the upper metastability limit of the melt. The lower metastability limit is between T_1 and some temperature T_2. Within this range of supercooling the centers are created rather slowly. Beyond the lower metastability limit the rate of formation of centers becomes immediately very high.

In the beginning of the development of the crystallization theory it was assumed that the formation of crystallization centers is induced by accidental (spontaneous) accumulations of atoms of the liquid which form a structure similar to the structure of a solid. If the size of these accumulations exceeds the critical size of the nucleus then, according to the theory, they are transformed into crystallization centers. This process is usually called the spontaneous or homogeneous nucleation of a new phase.

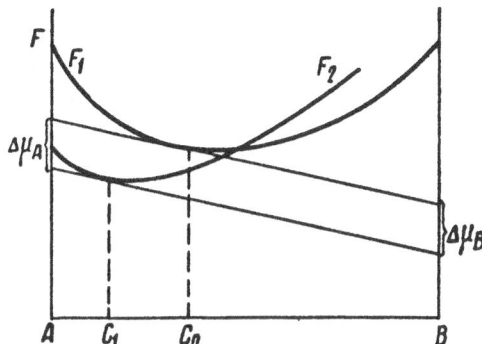

Fig. 24. Variation of the free energy
with the composition.

However, further studies of the nucleation of crystallization centers showed that the metastability limit is greatly dependent on the presence of impurities in the liquid and is often determined by the properties of the walls of the container in which the liquid is kept.

It was found that the metastability limit can change under the effect of an extremely small number of so-called active impurities which can be either soluble or insoluble.

From the theoretical point of view, the change in the supercooling temperature of the liquid as the result of adding a small amount of impurities can be due only to the decrease of the critical size of the nucleus.

If the impurity is soluble then the atoms must increase the force of atomic interactions and thus decrease the critical size of the nucleus. The impurity can also decrease the metastability limit without increasing the rate of formation of centers to any significant extent.

Insoluble impurities are usually present in the form of some dispersed compound with an already existing separation boundary. If the force of attraction of atoms of the liquid is sufficiently great at this boundary at a given supercooling temperature, then it facilitates the formation of nuclei because of the decrease in the total surface, which determines the work of formation of nuclei. In this case also, the fluctuational accumulation of some number of atoms with a structure of the solid must form at the surface of the insoluble impurity.

The nucleation of crystallization centers as the result of impurities is called heterogeneous nucleation.

The first investigations of the supercooling of unicomponent metal-like liquids were made at relatively low supercooling temperatures. As the experimental technique improved, the effect of extraneous factors was decreased and a considerable degree of supercooling was attained. A high degree of supercooling of liquids can be reached by essentially three different means: 1) by eliminating the impurities by zone melting and using appropriate slags and neutral crucibles; 2) by using very small volumes of the metal; and 3) by using high cooling rates.

In [91] 150 g of iron was supercooled 258°C under slag. In [92] iron purified by zone melting was supercooled 260-270°C.

In [93, 94] it was found that high degrees of supercooling can be attained in small drops of the melt. The method of investigating crystallization in microvolumes was improved and systematized in [95-97].

These investigations were based on the following principle. If the metal is divided into many drops (10^{-4}-10^{-9} cm^3 in size) then it is quite probable that some of the drops will not contain any atoms of the impurity. Therefore, some number of liquid drops will crystallize spontaneously. A great number of metals have been investigated by this method.

The third method is based on the principle that the metal can be cooled during a shorter period of time than is necessary for the formation of the first nuclei of the solid phase.

The supercooling reached in some pure metals is given here (°C):

Hg	77	Al	195	Ni	319
Ga	76	Ge	227	Co	330
Sn	118	Ag	227	Fe	295
Bi	90	Au	230	Pd	232
Pb	80	Cu	236	Pt	370
Sb	135	Mo	308		

The nucleation of crystallization centers in supercooled liquid solutions has been investigated very little. There are only a few data on supercooling of some binary systems [98,100]; data on the composition of nuclei are completely lacking. It is only from the data on the degree of liquation of crystals and solidified alloys that it is possible to ascribe a given composition to the nucleus. The analysis of this data and the theoretical data lead us to assume that the composition of the nucleus of the solid phase precipitated from the liquid solution must be determined by the equality of gradients of partial chemical potentials $\Delta\mu_A$ and $\Delta\mu_B$, at least in the case of high supercooling (below the metastability limit) [101]. Figure 24 shows the variation with concentration of the free energies $F_1(C)$ of the liquid solution and $F_2(C)$ of the solid solution. If the concentration of the alloy is C_0 then the composition of the precipitated nucleus is determined by the condition $\Delta\mu_A = \Delta\mu_B$, i.e., by the fact that the tangents ab and a_1b_1 are parallel. The point where the a_1b_1 is tangent to the curve of free energy of the solid solution must be close to the composition of the nucleus of the precipitated phase C_1. From the viewpoint of the molecular-kinetic theory, this is explained by the fact that the forces of atomic interaction are different from zero only at relatively small distances, and consequently the atoms directly adjacent to the boundary of the nucleus must unite with the nucleus under the effect of the same forces. If we imagine that great numbers of A atoms unite with the nucleus then $\Delta\mu_B$ becomes greater than $\Delta\mu_A$ and the forces acting on the B atoms increase. At the following moment the concentration changes and becomes an equilibrium concentration. This process by which the equilibrium concentration is maintained occurs in a very small volume next to the boundary zone. Therefore, it occurs over extremely short periods of time, of the order of the period of a few oscillations of the atom. All this discussion is valid only after the nucleus is already formed, i.e., after the creation of the interphase separation boundary. One can say that these processes correct the composition of the nucleus during its growth rather than determine its concentration at the moment of formation. Nevertheless, the process of fluctuation itself, i.e., the formation of the nucleus, is not the most important factor; rather, it is the "survival" time of the nuclei. In other words, nuclei of any composition (more or less close to C_1) can be created as the result of fluctuation, but the nuclei having a composition C_1 will have the greatest survival because they can grow without losing time on different oscillations resulting from deviations from the concentration C_1, at which the forces acting on the different kinds of atoms are in equilibrium. In what follows, the state in which $\Delta\mu_A = \Delta\mu_B > 0$ will be called the quasi-equilibrium state.

The supercooling of metal-like liquids is presently one of the most obscure problems of the theory of the liquid state. Perhaps the most important part of this problem is the relationship between fluctuations and the formation of crystallization centers. The fact is that density fluctuations in a unicomponent system and concentration fluctuations in liquid solutions may also occur at temperatures above the solidification temperature of the alloy [102]. These fluctuations also exist in supercooled liquids; the size of the fluctuations can apparently exceed the critical size of the nucleus, which is determined by Eq. (5.3). The principal difference between a fluctuation and a nucleus is that the nucleus has a separation boundary with the surrounding phase.

Thus, the formation of nuclei must be related not to the density or concentration fluctuations but to the probability of formation of a clear interphase separation boundary. In this light, the role of impurities, walls, and different external forces such as ultrasonic fields is very important.

In technically pure alloys crystallization centers usually nucleate on impurities. Therefore, in practice it is difficult to find a strict relationship between the rate of formation of nuclei and supercooling. This is even more difficult when the transformation occurs in the solid state [103]. In this case the decisive role is played by the different types of defects, heterogeneity, and the already existing separation boundary between crystallites of the mother phase [104]. Volume changes, orientational relationships, etc., play important roles in the precipitation of the nuclei of the new phase.

All these details require corrections in the classical equation for the rate of nucleation of crystallization centers.

In its present state the theory gives relatively precise data on the work of formation of nuclei and the supercooling of the alloy. Thus, the theory describes the kinetics of the formation of centers during the early stages of nucleation. This is usually used to determine the surface tension at the nuclear-mother phase boundary, the value of which has been determined by indirect methods up to now. The order of magnitude of this value determines the quantitative agreement between the theoretical and experimental data. In any case, the shapes of the experimental and theoretical curves $I(\Delta T)$ are almost always identical.

For a given degree of supercooling, the number of centers formed is determined first of all by the value of the surface tension at the nucleus-mother phase boundary. The surface tension depends on the temperature and on the singularities of the structure of the interphase separation boundary.

In the case of the martensitic transformation, the nucleation of crystallization centers has a singular character. Martensitic nuclei occur almost instantaneously when a given degree of supercooling is reached. Their number is independent of the time the steel is kept at this temperature and is determined for different steels only by the degree of supercooling and the singularities of the sample.

A certain critical cooling rate is necessary for the formation of martensite crystals. It should be emphasized that martensite nuclei can form only at temperatures at which no concentration fluctuations are possible.

The complexity of the mechanism of this transformation has made it impossible thus far to generalize the large number of experimental data obtained by Kurdyumov, Shteinberg, and their students and followers.

The characteristic trait of the nucleation of crystallization centers in the solid state is the preferential precipitation of a new phase at the boundary between grains of the mother phase or on crystallographic planes which are far removed from each other.

In concluding this section, we should note that direct observation of the nucleation process of crystallization centers has been impossible because of their small size (radius close to 10^{-7} cm). All the experimental methods of studying this phenomenon require in one way or another some period of time in which the nucleus has time to grow to a certain size before being detected. The structure of the alloy is determined by nuclei already grown to some size and having well-defined separation boundaries. The formation of nuclei of a critical size and their transformation into crystallization centers (i.e., into nuclei capable of growth) is also related to the kinetics of the process by which atoms combine with the nuclei. Therefore, the nucleation and growth of crystals are governed by the same laws. During the nucleation and growth of nuclei there is a separation boundary, and this in itself indicates that the forces acting on the atoms in both phases remain in equilibrium at the boundary.

In the equilibrium state $\Delta \mu_A = \Delta \mu_B = 0$, while during growth $\Delta \mu_A = \Delta \mu_B > 0$. If this condition were not satisfied then the boundary would be very much widened and the nucleus would resemble a fluctuation.

2. Crystal Growth

A great many investigations have been made of crystal growth. During the past three decades several magnificent monographs on different problems in crystal growth have appeared [90, 105-108] and in recent years there have been several books on the mechanism of crystal growth [109, 110]. In addition, a great number of reviews describing more or less completely the modern state of the problem have been published [111-113]. However, in spite of all these works, several points in the process of crystal growth remain unexplained and much time and effort will be needed to clarify them.

According to the laws established by Tammann, the rate of crystal growth* must increase with the degree of supercooling until the viscosity of the liquid becomes so great that the molecules can no longer pass from the liquid to the solid state. In metastable liquids the rate of crystal growth usually increases up to a considerable degree of supercooling [114].

The earliest stages of the growth of nuclei are the most important in the process of solidification of alloys.

However, most experimental data on the rate of crystal growth refer to the later stages of growth. The growth rate has most often been studied by using seed crystals, i.e., very small crystallites introduced into the liquid in order to measure the growth rate in different directions.

Heat is being evolved at the crystallization front during the time that the atoms join the growing crystal. This evolved heat can reduce supercooling to zero by heating the liquid adjacent to the crystal.

Thus, continuous crystal growth requires a temperature gradient. The heat can be conducted through the liquid or through the solid phase. In the first case the temperature of the liquid must be below the temperature

---
* In what follows we shall use the expressions growth rate and solution rate to mean the rate of motion of the interphase separation boundary in a given direction.

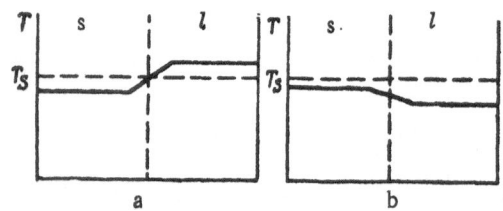

Fig. 25. Distribution of the temperature at the crystallization front. a) Conduction of heat through the solid phase; b) conduction of heat through the liquid phase.

of the crystal and in the second case the temperature gradient must have an opposite sign. However, the temperature of the liquid far away from the crystal can be even higher than the melting point. Naturally, in both cases the temperature of the crystal must be below the melting point. The temperature gradient at the crystallization front is determined by many factors, e.g., the degree of supercooling, the crystallization heat, the method of heat conduction (through the crystal or through the liquid), heat conductivity, heat capacity, convection, viscosity, etc. The distribution of the temperatures at the crystallization front is represented schematically in Fig. 25.

If the crystal grows from a supersaturated solution, then the growth rate is determined by the concentration gradient at the crystal-liquid boundary.

Studies of crystal growth with seed crystals have shown that in some cases crystals grow when the degree of supercooling is no more than 0.1°C, while in other cases they being to grow only when the degree of supercooling is rather high.

The earliest investigations of crystal growth from transparent liquids showed that a diffusional layer is formed around the growing crystal and that there is a concentration gradient in this diffusional layer [117, 118]. The thickness of this layer can vary within considerable limits. If the liquid is intensely mixed the thickness of the layer reaches a molecular size. These layers are sometimes called "crystallization enclosures." In a unicomponent system the growing crystal may be surrounded with a rather thin thermal jacket, particularly if the heat is conducted away through a solid phase or when the growth rate is very high.

There have been very few studies of crystal growth in the case of two-phase decomposition of supersaturated liquid solutions. There is one very good monograph, unique in its way, by Bochvar [115].

The study of the crystallization of liquids with a eutectic composition [115, 116, 119, 120] showed that eutectics can solidify either by separate crystallizations or by eutectic colonies.

If the growth of one of the phases has no nucleating effect on the second phase, then the second phase can nucleate far away from the first phase and the two phases can crystallize independently. This phenomenon is particularly characteristic in the case of low degrees of supercooling [115]. Purely eutectic crystallization occurs only when the nucleation of one phase (leading phase) at some stage of growth leads to such an enrichment of the surrounding layer of liquid with the second component that crystallization centers of the second phase occur in this layer of liquid. (Such crystallization centers can also occur away from the growing nuclei of the leading phase. Here, the fact that they are touching is important.)

When there is simultaneous growth of both phases the growth rates are much higher, since diffusional distribution of atoms occurs within a narrow crystallization zone directly adjacent to the crystallization front. As a result, a eutectic colony is formed in which one of the phases is the leading phase. Very often this phase grows in the form of cylindrical or oval threads which are thicker in places because of temperature and concentration fluctuations. The second phase usually envelops the threadlike crystals of the leading phase.

The growth of separate crystals during phase transformation in a solid state has many singularities. Nuclei of the second phase grow in the mother phase with a certain anisotropy of properties. In many cases the density of the precipitating phase is very different from the density of the initial solid phase. The precipitation and growth of the second phase is accompanied by the formation of a stress field which in many cases handicaps further growth. If further growth is not handicapped, then vacancies occur in the surrounding matrix simultaneously with the growth of the new phase. These vacancies diffuse into the matrix from the crystallization front.

In the case of two-phase decomposition of a supersaturated solid solution (eutectoid decomposition) the distribution of the substance at the front of the growing colony remains essentially the same as in the case of eutectic crystallization. However, in the solid state the growth of the colony is controlled by the condition that the orientations be the same, since the initial phase has a specific crystal lattice. Eutectoid colonies usually consist of a series of plates spaced at a rather precise distance, which is determined by the degree of supercooling [41]. Experimental results show that the growth rate of a pearlitic colony is constant when the temperature of the eutectoid decomposition of austenite is constant [121].

Unfortunately, the data on the growth rates of crystals are spotty and not systematized. This is particularly true in the case of crystal growth in metals during phase transformation because of the great experimental difficulties in determining the growth rate of crystals under these conditions. The few measurements providing dependable data on the growth rate of crystals refer mainly to dendrites. Some investigations have been made of the early stages of annealing of martensite. In particular, the authors in [140] determined the average size of carbides during annealing of martensite. Konobeevski[204] has done considerable work on the crystal growth of particles of the dispersed phase during aging of alloys. He showed that when the rate of formation of nuclei is high, the growth of crystals of the dispersed phase may rapidly stop because of the so-called colloidal equilibrium. This is an equilibrium between the dispersed particles and the solid solution having a much higher concentration than the equilibrium concentration corresponding to the phase diagram. The later and much slower growth of crystals can occur only as the result of coalescence.

Apparently, the most complex type of crystal growth is that of martensite crystals. These crystals grow very large at a tremendous rate at temperatures close to absolute zero. Attempts to measure the growth rate of these crystals have not succeeded. In [124] the author succeeded in showing that the rate of displacement of the edge of the martensite plates in an alloy of iron containing 29.5% Ni is about 10^5 cm/sec. In this alloy the growth rate of martensite crystals remains constant at temperatures between -20 and $-195°$C.

The theory of crystal growth has developed in two directions. In the early studies the growth rate of crystals was calculated by using the transport equation and taking different transformation conditions into account.

Later investigations have been based on a definite model of the structure of a crystal, in particular the model of the structure of the surface of separation of phases. To determine the growth rate one usually calculates the probability of the joining of some number of atoms to the surface of the crystal.

In the earliest work on crystallization [122-124] crystal growth was considered to be a purely diffusional process. The amount of substance passing from the solution into the crystal dq during time dt was determined by

$$dq = D \frac{(C_0 - C_\sim)}{x} S dt,$$

where D is the diffusion coefficient; S is the area of the surface of the crystal; $C_\sim$ is the concentration of the dissolved substance at the saturation point; C_0 is the concentration of the supersaturated solution; and x is the thickness of the layer through which diffusion occurs. In the case of intense mixing the value of x can be very small [118].

Further development of the theory was based on more precise calculations resulting from the solution of the equations of diffusion for more complex boundary conditions.

Thus, Frenkel' showed that the growth rate of a spherical nucleus of a phase precipitating during isothermal decomposition of a supersaturated solution is determined by the following relationship [77]:

$$v = \frac{dr}{dt} = \frac{D}{\rho r} \left[(C_1 - C_\sim) - \frac{a}{r} \right], \tag{5.6}$$

where $\Delta C = C_1 - C_\sim$ is the difference between the concentrations of the supersaturated and equilibrium solutions; r is the radius of the growing nucleus; and a is a magnitude entering the Thompson equation (3.5): $a = 2\sigma M C_\sim / \rho RT$.

The shape of this function $[v(r)]$ is shown in Fig. 26. The growth rate reaches a maximum when the radius of the growing crystal is twice the critical size of the nucleus. From Eq. (5.6) it is easy to find the dependence of the growth rate of the crystallization centers on the time t during which the crystals are at constant temperature during the later stages of growth:

$$v = \sqrt{\frac{D\Delta C}{2\varrho t}}.$$ (5.7)

More complex calculations taking into account the nonsteady state of the process were made in [125, 126].

Somewhat later, the growth rate of separate crystals during the late stages of the decomposition of a supersaturated solution was determined by taking into account the stresses occurring at the front of the growing grain [127, 128].

The growth rate of pearlite colonies and the solution of the diffusion problem were given in [129-131]. The theoretical and experimental data obtained in [121, 134] agree satisfactorily.

The growth rate of martensite crystals has not been determined theoretically except in [135]. In this work the author explains the change in the size of martensitic grains with the temperature observed in [136].

Since a rather high degree of supersaturation is necessary for the growth of perfect crystals, a detailed investigation must be made of the mechanism by which the atoms of the original phase combine with the surface of the growing crystal.

Calculations showed that this phenomenon can occur if the atoms combine with the surface of the crystal not by sticking of separate atoms but as the result of the deformation of the so-called two-dimensional nuclei [132,133].

A two-dimensional nucleus is a monoatomic layer of a size which ensures its stability at the boundary with the mother phase.

The critical size of the two-dimensional nucleus is determined in the same way as the size of the three-dimensional nucleus. The work of formation of such a nucleus is

$$F = \frac{1}{2}\sigma'l,$$

where σ' is the free energy per unit length of the steps and l is the perimeter of the nucleus.

When a two-dimensional nucleus is formed, the surface of the crystal is rapidly covered with a new layer of atoms. The growth rate of the crystal is determined by the probability of the formation of two-dimensional nuclei and is expressed by a relationship similar to (5.5).

The crystal growth resulting from the formation of two-dimensional nuclei requires a high degree of supercooling. In most cases, particularly in alloys, the crystals grow when the degree of supercooling is very low.

Because of this result the growth mechanism must be corrected by taking into account the imperfections in the crystal—screw dislocations [137]. Dislocations facilitate the atoms joining the surface of the crystal [113].

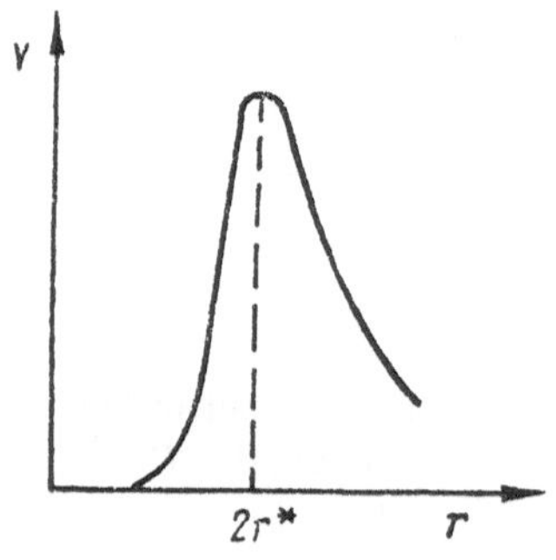

If the rate of formation I and the growth rate v of crystallization centers are known then in the simplest case the time during which an alloy becomes crystallized is determined by the following relationship [276]:

$$t = \frac{K}{\sqrt[4]{Iv^3}},$$

where K ~ 1.

Fig. 26. Rate of crystal growth as a function of the radius of the crystal, according to [5, 6].

Let us investigate in more detail the mechanism of the growth of crystals.

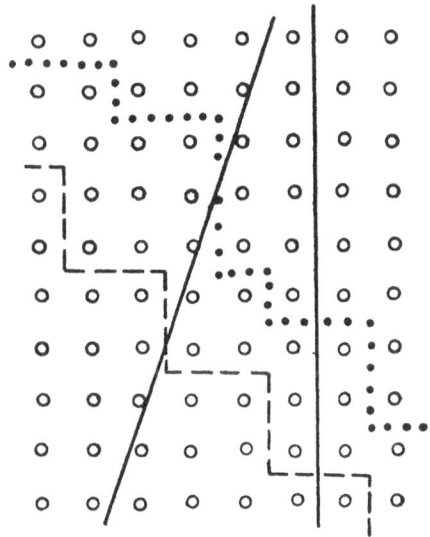

Fig. 27. Smooth and stepped phases [253].

Numerous observations have shown that crystals can acquire very different shapes, facetings, and habits* under different growing conditions. In all types of growth the gain of the free energy ΔF during the transformation of the crystallizing phase (I) into the solid phase (II) must exceed the surface energy resulting from the growth of the total separation boundary ($\Sigma \sigma_i S_i$). We shall consider two limit cases.

In the first case growth occurs under conditions close to equilibrium conditions (equilibrium growth). In this case ΔF is small, and consequently the shape of the crystal must satisfy the Gibbs-Curie-Wolff principle. Under these conditions perfect single crystals with plane, mirror-like faces can usually be grown.

In the second case, crystals grow under conditions very different from equilibrium conditions. In practice, such conditions are quite common.

Under conditions very different from equilibrium conditions (labile state) the total free energy of the surface of separation increases as the result of the occurrence of faces with high values of surface tension, as the result of irregularities (rounding off) or the increase of the total length of the separation boundary resulting from the growth of branched crystals. In all cases the different areas of the boundary of the crystal can be either mirrorlike or rough faces (steplike) [278]. These faces are shown in Fig. 27. Rounded crystals also have a steplike boundary, where the height and frequency of steps determine the imaginary curvature of the separation boundary.

The concept of the mechanism of crystal growth has been developed from a large number of theoretical and indirect experimental data.

In this concept the details relative to the formation of different types of crystal growth are quite unclear. The first serious attempts to clarify these details can be found in [278, 280].

In general terms, the mechanism and subsequent stages of growth of a crystal are represented schematically in Fig. 28.

If a crystal with smooth faces is surrounded by the supercooled phase (II) its faces begin to grow as the result of the formation of a two-dimensional nucleus. Then the atoms adjoin the corners formed by the two-dimensional nucleus and the face itself. Thus, layer after layer the faces grow under conditions close to equilibrium conditions.

Dynamic equilibrium is continuously established at the boundary of separation between the crystal and the surrounding medium. Every act through which atoms or groups of atoms unite with the crystal leads to the evolution of crystallization heat; this heat is spent on the increase of the kinetic energy of atoms immediately adjacent to the crystallization front. In the case of a low degree of supercooling, when the atoms unite with the crystal very slowly, the evolved crystallization heat has time to be conducted away from the crystallization front before any significant heating occurs.

An increased degree of supercooling results in a much higher rate of nucleation of two dimensional nuclei and subsequent nucleation of faces. If, at the same time, the conductivity of the surrounding medium is not too high, then the crystallization front may become locally overheated and the crystal may begin to melt. In this case not all the areas of the crystal grow at the same rate. Figure 28 shows two of the many possible types of crystal growth.

Figure 28a shows the case where two-dimensional nuclei occur in different areas of the crystal surface and grow toward each other. In this case the maximum heat is evolved between the growing layers, and consequently

---
* Habit is the ratio between the sizes of the different faces of a crystal.

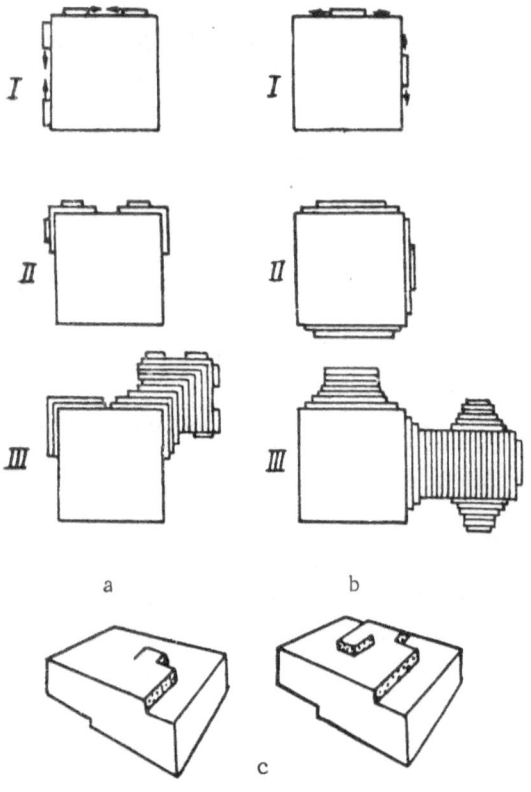

Fig. 28. Growth mechanism of crystals according to [264, 285]. a,b) Two-dimensional nuclei; c) screw dislocation.

the growth of the whole face is slowed down. At the next moment a nucleus is formed somewhat away from the corner because phase II at the corner is still overheated. During the growth of the first layer over the crystal faces new two-dimensional nuclei are created. These begin to grow and add some crystallization heat to the area between the nuclei growing toward each other. Crystal growth continues in this sequence.

Figure 28a shows that the apexes are the most advantageous directions of crystal growth. The boundary faceting becomes stepped. When the protrusions of the growing crystal become sufficiently long so that they no longer affect each other,the ends of the protrusions begin to thicken. When a certain thickness is reached, two-dimensional nuclei are created at the end of the protrusions and these, in turn, produce new protrusions. The skeletons of crystals are thus formed. Figure 28b shows a few successive stages of the formation of protrusions in the case where the two-dimensional nuclei occurring in the center of faces grow rapidly. The growth rate of such protrusions must be very high because the heat at the crystallization front is conducted through a spherical surface which is larger than the surface of the end of the growing protrusion.

In many cases, branched crystals are formed when the degree of supercooling is low, a condition in which two-dimensional nuclei cannot form. The most probable growth mechanism in this case is growth by intermediate screw dislocations (Fig. 28c). Screw dislocations form steps at the surface of the crystal and the crystal grows because the atoms unite with these steps. Two stages of the growth of a crystal by this mechanism are shown in this figure.

Under the particular conditions of electrocrystallization or crystallization from vapor this type of growth may lead to the formation of threadlike crystals (whiskers) [23, 281, 282]. These crystals have almost no moving dislocations and have an anomalous high strength.

If the transformation occurs in a unicomponent system then the growth rate of the crystal is completely determined by the heat conductivity and the mechanism by which the atoms unite with the surface of the growing crystal.

In the case of the decomposition of supersaturated solutions the growth rate is determined also by the diffusion of the substance composing the crystal to the crystallization front. In this case the thermal and diffusional flows interact and create conditions favorable for the growth of smooth or stepped faces.

The stepped surfaces can be enscribed in a sphere formed by the temperature field and thus give the impression of a rounded separation boundary.

Surface tension at the place boundary plays an important role in crystal growth. The surface tension is the sum of the surface free energy of plane areas, the free energy of the edges, and the free energy of the corners. If the curvature of the separation boundary is large then the free energy of the edges and the corners increases. At the boundary with a small curvature the surface tension is essentially the sum of the surface tensions of the planes. Clearly, the total gain of the free energy of transformation is greater if the growing crystal is bound by surfaces with low surface tension. The general theory of crystal growth indicates that the formation of two-dimensional nuclei and of the steps of screw dislocations is the most probable on crystal faces with low surface tension. On the basis of this fact and of the mechanism of crystal growth described here it follows that the motion of the crystallization front is accompanied by an increase in the curvature of the separation boundary, the surface tension varying very little.

A convex stepped surface with planes having low surface tension is formed during the growth of branched crystals. Thus, the free energy of the system increases not as the result of surface tension but as the result of the increase of the total length of the separation boundary.

3. Different Types of Crystal Growth

Let us examine the general relationships determining the different types of crystal growth [138, 139]. These relationships must be general because the number of different phases and transformation conditions is for practical purposes infinitely large, while the types of crystal growth have only a few fundamental differences.

The analysis of a large number of experimental data concerning crystal growth shows that there are three mechanisms of crystal growth.

1. The plane boundary of separation between the crystals and the surrounding mother phase remains unchanged during crystal growth.

2. The convex separation boundary with a given radius of curvature remains unchanged during crystal growth.

3. The radius of curvature of the interphase boundary increases, while the crystal grows from the convex surface of separation.

In more complex cases the three mechanisms may occur simultaneously.

Crystal with plane faces grow according to the first mechanism. The perfect single crystals used in the optical industry grow by this first mechanism.

Skeleton crystals — branches of dendrites, spherulites, needles, and plates with convex separation boundaries (the convexity being in the direction of high growth rates) — grow according to the second mechanism.

All rounded crystals grow according to the third mechanism. This type of growth occurs most often during the later stages of the decomposition of supersaturated solutions or when the degree of supercooling is insignificant, the liquid is not thoroughly mixed, etc.

General considerations of the molecular-kinetic mechanism of phase transformation lead us to assume that crystals grow by the mechanism which ensures the maximum rate of displacement of the crystallization front into the mother phase (in a given direction).

In fact, the rate of transfer of atoms from one phase to another is determined by the mobility of the atoms and the magnitude of the forces on the atoms.

The magnitude of these forces acting on atoms of a given kind depends on the magnitude of the partial chemical potential gradient. Therefore, if there is in some direction a configuration of the field of thermodynamic activity such as to ensure the maximum rate of combination of atoms with the crystal then the rate of growth of crystals in this direction will be greater than in other directions until some accident disturbs this favorable circumstance. The configuration of the field of thermodynamic activity surrounding the growing crystal depends on many factors. Among these factors, the first is supersaturation and supercooling of the initial phase, since it is only when the phase is supersaturated or supercooled that the forces promoting the transformation of one phase into another can occur. We refer here to supersaturation and supercooling with respect to complete equilibrium determined by a plane phase separation boundary. However, if the nucleus grows then the concentration of the surrounding solution is somewhat higher than the equilibrium concentration at the deformed surface as the result of the same dynamic equilibrium.

The heat of crystallization affects the distribution of the concentration at the crystallization front and consequently affects the temperature distribution upon which, in the final analysis, the diffusion coefficient itself depends.

The surface tension at the phase separation boundary is very important; the surface tension can be different in different directions. It is clear that if one takes into account all these factors, the problem becomes very complicated and the solution will be difficult to analyze. Therefore, we shall simplify the problem by introducing a magnitude which will allow us to neglect a number of factors. This magnitude is the thickness of the crystallization enclosure. There is no doubt of the existence of a crystallization enclosure, since the crystallization centers cannot change the concentration of the surrounding solution at any great distance from the crystallization front. The concentration changes over a narrow region which is determined by the thickness of the crystallization enclosure, and the thickness of this enclosure itself depends on the many factors previously mentioned. When a steady state is established, the thickness of the crystallization enclosure (the change in its thickness) is determined by the ratio between these continuously acting factors.

Thus, the problem is reduced to determining the growth rate of a spherical nucleus of the phase precipitated from the supersaturated solution. During the growth of the nucleus (Fig. 29a) a layer of the mother phase impoverished in the precipitating substance is formed around the nucleus, and the substance diffuses through this layer. To simplify the problem we shall assume that the nucleus consists of a pure substance and that the amount of the solution is so large that its composition does not change during the growth of the nucleus. If we represent the crystallization enclosure as a hollow sphere with an inner diameter r and outer diameter r_1 (r being simultaneously the radius of the growing nucleus) then, according to the diffusion theory, we have the following relationship for the diffusional flow with spherical symmetry in the steady state:

$$\frac{1}{r_2} \cdot \frac{\partial^2 (r_2 C')}{\partial r_2^2} = 0.$$

For the diffusion through the hollow sphere with inner and outer radii r and r_1 the boundary conditions are respectively $C' = C_1$ when $r_2 = r$, regardless of the value of t, and $C' = C$ when $r_2 = r_1$, regardless of the value of t.

With these limit conditions the solution of the diffusion equation is

$$\frac{C_2 - C'}{C_2 - C} = \frac{r_1}{r_2} \cdot \frac{(r_2 - r)}{(r_1 - r)},$$

where C' is the concentration in the steady state for any value of r_2.

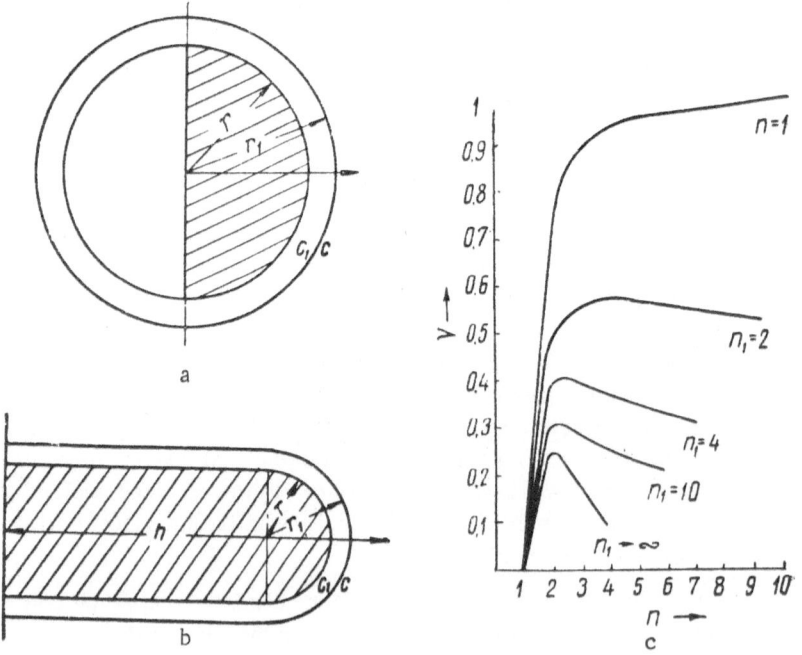

Fig. 29. Calculation of the growth rate. a, b) Diagram used in the calculations;
c) growth rate at different values of x.

The flow of the substance per unit surface of the sphere per unit time is

$$j = -D \left(\frac{\partial C'}{\partial r_2} \right)_{r_2 = r_1} = -D (C_1 - C) \frac{r}{r_1 (r_1 - r)} .$$

For a hollow sphere the expression takes the form

$$Q = -4\pi r_2^2 D \left(\frac{\partial C'}{\partial r_2} \right)_{r_2 = \rho} = -4\pi D (C_1 - C) \frac{r r_2}{r_2 - r} .$$

The amount of the substance diffused during the time t is

$$q = 4\pi D \int_0^t \frac{(C_1 - C) r r_2}{r_2 - r} \, dt,$$

where C is the concentration of the solution at the outer boundary; C_1 is the concentration immediately at the boundary of the growing nucleus; the value of C_1 depends on the curvature of the separation boundary and can be expressed as a function of r by the approximate Thompson equation:

$$C_1 = C_\sim \left(1 + \frac{2\sigma M V}{RTr} \right). \tag{5.8}$$

Then the diffusion rate of the substance through the impoverished zone can be represented as

$$\frac{dq}{dt} = 4\pi D \left(\Delta - \frac{2\sigma M C_\sim V}{RTr} \right) \frac{r r_2}{r_2 - r} ,$$

where $\Delta = (C - C_\sim)$ is the degree of supersaturation of the solution and V is the specific volume of the precipitating phase.

This equation can be used to determine the linear growth rate of crystallization centers if we express q in terms of volume

$$q = \tfrac{4}{3} \frac{\pi r^3}{V}, \text{ and } dq = \frac{4\pi r^2}{V} \, dr.$$

Then the growth rate becomes

$$v = \frac{dr}{dt} = DV \left(\Delta - \frac{a}{r} \right) \left(\frac{1}{x} + \frac{1}{r} \right), \tag{5.9}$$

where x is the thickness of the impoverished zone.

Integration of this expression gives the dependence of the crystal size on the time of isothermal growth. However, to analyze the process of formation of different shapes of growing crystals it is more convenient to use Eq. (5.9), which gives the dependence of the rate on values having a simple physical meaning.

If we represent graphically the dependence of the growth rate on the crystal size for different values of the thickness of the crystallization enclosure we obtain the family of curves shown in Fig. 29c.

The shape of the growing crystal is determined by the conditions under which the crystallization center develops and by the tendency of the growth rate toward a maximum value.

The behavior of the crystallization enclosure during crystal growth is the consequence of the interaction of a number of factors. These factors are diffusion, evolved crystallization heat, convection, accumulation of impurities at the crystallization front, stress, plastic deformation, etc.

All these factors determine the thickness of the crystallization enclosure to some degree. The conditions of growth can be divided into two different combinations of factors according to the behavior of the impoverished zone.

The first is the combination of factors occurring when the thickness of the impoverished zone grows with the crystal; the second is the combination of factors occurring when the thickness of this zone remains constant regardless of the size of the growing crystal.

We shall show below that these conditions and the crystalline structure of the growing center determine the shape of the crystal.

Up to now, no general relationship has been found which would make it possible to determine how the thickness of the zone changes with different growing conditions. At present the changes can be deduced only from indirect data concerning relatively simple growing conditions.

On the basis of this discussion we shall describe some characteristic cases of crystallization from supersaturated solutions.

1. The crystallization center precipitates from the supersaturated solution of the phase and grows in the direction indicated by the arrow in Fig. 29a. As the result, the thickness of the impoverished zone increases as the crystal grows. This growth mechanism occurs most often in the case of decomposition of slightly supersaturated solid solutions. Since the atoms of the solvent are fixed at the sites of the crystal lattice and the atoms of the precipitating phase move toward the crystallization front, the impoverished zone grows during crystal growth because the structural material from the solid solution arrives from increasingly greater distances.

In alloys this mechanism occurs in the case of growth of cementite centers during the decomposition of martensite during aging, etc.

Let us determine the value of x when the composition of the growing center remains constant during growth. In this case the amount of substance precipitated from the impoverished zone is

$$q = \frac{2}{3} \pi (r_1^3 - r^3)(C - C_{av}),$$

where C_{av} is the average concentration of the precipitating substance in the impoverished zone.

The part of the crystallization center whose growth we are investigating consists of the same amount of substance:

$$q = \frac{2}{3} \cdot \frac{\pi r^3}{V}.$$

From these two equations one easily obtains the thickness of the impoverished layer

$$x = r_1 - r = r\left[\sqrt[3]{\frac{1}{V(C - C_{av})} + 1} - 1\right].$$

If we substitute this value into (5.9) we obtain

$$v = DV\left(\frac{\Delta}{r} - \frac{a}{r^2}\right)B,$$

where

$$B = \frac{\sqrt[3]{\dfrac{1}{V(C - C_{av})} + 1}}{\sqrt[3]{\dfrac{1}{V(C - C_{av})} + 1} - 1}.$$

When the degree of supersaturation is low then $B \sim 1$ and

$$v = DV\left(\frac{\Delta}{r} - \frac{a}{r^2}\right). \tag{5.10}$$

This equation shows that the rate of crystal growth first increases rapidly and then decreases as the radius increases. It can be shown that the growth rate is maximum when $r = 2r^*$, where r^* is the radius of the critical nucleus.

In fact, it follows from (5.8) that $r = a/(C - C_\sim)$. When the size of the nucleus is critical the concentration of the mother phase at the boundary with the nucleus is the same as that of the surrounding supersaturated solution, i.e., $C_1 = C$, (since the critical size of the nucleus is the center of the precipitating phase having a size which is in equilibrium with the surrounding supersaturated solution). It follows then that $r^* = a/(C - C_\sim)$.

If we substitute $r = nr^*$ into (5.5) (where n is the number of critical radii contained in r) we obtain

$$v = \frac{DV\Delta^2}{a} \left(\frac{n-1}{n^2} \right).$$

This function has a maximum for $n = 2$.

The result obtained is in agreement with the results obtained by Frenkel' and Pines in a somewhat different way.

One of the most important consequences of Eq. (5.10) is that the surface tension σ entering in the constant a affects the growth rate only during the initial stages of growth, since the second term in the expression within the brackets can be neglected for later stages. Since the surface tension is different in different directions, crystals precipitate during the first stages of growth either as plates or needles and then are transformed into equiaxial crystallites. Thus, when martensite decomposes, the cementite crystals have the shape of discs in the beginning and only later are transformed into globules [140]. This probably occurs because the rhombohedral lattice of cementite has three directions in which the surface tension is minimum, and consequently the growth rate perpendicular to the $\{100\}$ faces is maximum, as one can judge from the density of iron atoms at the faces. The maximum growth rate is in the direction perpendicular to the $\{100\}$ face – the growth rate is somewhat less in the direction of the $\{010\}$ face. In the other directions the growth rate is even lower. The difference in the growth rates in different directions during the early stages, when the crystal is still small, leads to the formation of platelets. When the crystals reach a size at which the surface tension ceases to be important the shorter parts of the crystal grow faster than the longer parts because the growth rate depends on the size of the crystal. Thus, in this stage of growth the platelets are transformed into equiaxial crystals.

This scheme is, of course, imprecise. For a precise picture of this process one must also take into account the singularities of the surrounding solid solution.

In most cases the precipitation from the solid solution occurs either at the boundary between the crystallites of the mother phase or between the planes with low indices – most often at the boundaries of mosaic blocks. However, there are always one or two directions in which the surface tension at the phase boundaries is low, and this direction determines the shape of the growing crystal.

When the degree of supersaturation is low and the thickness of the zone increases together with the growth of crystallization centers the most advantageous shape is a sphere in the later stages of growth. This is clear from the direct analysis of Eq. (5.10) and also from the experiments made by Arbuzov.

2. Let us investigate the case where crystallization centers grow under conditions such that from a given moment the thickness of the impoverished layer x remains constant. These growing conditions occur most often during the decomposition of supersaturated liquid solutions. In many liquids the mobility of the molecules is so high that the outer boundary of the impoverished zone is continuously destroyed because of the intense mixing of the molecules in the supersaturated mother phase. Some constant thickness of the impoverished zone is established during the process of isothermal growth. This thickness is determined essentially by the diffusion rate of the molecules of the precipitating phase, the total mobility of the molecules of the mother phase, and the growth rate of the crystal itself.

When the growth rate is high, particularly when the volume increases during crystallization, then crystals can grow into the impoverished zone. In this case the thickness of the impoverished zone remains constant because the crystal growing into the impoverished zone pushes it to either side and grows towards its outer boundary.

This is the characteristic mechanism in more viscous liquids. Quite often the thickness of the impoverished zone remains constant during crystal growth in the case of the decomposition of solid solutions. In many cases the decisive role is played by the degree of supercooling resulting from certain conditions of heat conductivity. Cementite plates grow in this manner in iron-carbon alloys during the formation of the Widmanstatten structure in posteutectoid steels. A somewhat different mechanism of the formation of an impoverished zone with a constant thickness occurs during the growth of pearlite colonies during eutectoid decomposition of austenite. In this case the principal concentration gradient determining the growth rate occurs between the crystallization fronts of the cementite and ferrite plates. The distance between the plates remains constant, and this is responsible for the constant thickness of the impoverished zone.

The variation of the rate of crystal growth at constant thickness of the impoverished zone (x = const) can be derived from Eq. (5.9). If we express the radius of the crystal and the thickness of the zone in terms of the radius of the critical nucleus

$$x = n_1 r^*, \quad r = nr^*$$

and we use the relationship $r^* = a/\Delta$, we can rewrite Eq. (5.9) as follows:

$$v = \frac{DV\Delta^2}{a}\left(1 - \frac{1}{n}\right)\left(\frac{1}{n_1} - \frac{1}{n}\right).$$

The variation of the growth rate with the size of the growing crystal is represented in Fig. 29c, where each curve corresponds to a given value of n_1.

The curves show that the growth rate increases with decreasing thickness of the impoverished zone and each curve relative to $x > r^*$ has a maximum growth rate which is displaced toward greater sizes of the growing crystal with decreasing thickness of the impoverished zone. When the value of x is small the maximum disappears almost completely.

This analysis concerns the growth of a spherical crystallization center. To find out what shape of the growing crystal is the most advantageous from the energy standpoint when the thickness of the impoverished zone remains constant, we shall examine two limit cases.

1. The thickness of the impoverished zone $x \ll r$. Then we can neglect $1/r$ as compared to $1/x$ in Eq. (5.9); as the result we obtain

$$v = \frac{DV}{x}\left(\Delta - \frac{a}{r}\right). \tag{5.11}$$

This relationship shows that the growth rate of crystals increases with increasing radius of curvature of the phase separation boundary. One can see that in this case the maximum growth rate corresponds to the crystals growing in the form of polyhedrons with plane faces. This conclusion is in agreement with the experimental data which show that crystals in the form of polyhedrons with plane faces are formed mostly during the decomposition of liquid solutions with low viscosity or in the case of crystallization from a gaseous medium.

2. The thickness of the impoverished zone x is constant but greater than r. When x is constant the crystals grow with a constant radius of curvature of the separation boundary.

In this case the growth of crystals in the shape of a cylinder of height h terminated by a half sphere of radius r with the convexity in the direction of growth (Fig. 29c) is the most characteristic.

Then

$$q = \pi r^2 \frac{h}{V},$$

and

$$v = \frac{dh}{dt} = 2DV\left(\Delta - \frac{a}{r}\right)\left(\frac{1}{x} + \frac{1}{r}\right).$$

When x is large the size of the radius for which the growth rate is maximum is equal to twice the radius of the critical nucleus. When the thickness of the impoverished zone is smaller, the radius of curvature and the thickness of the cylinder for which the growth rate is maximum increase. Thus, for each value of x the thickness of the growing crystal remains constant and ensures the maximum growth rate, which can be determined from the curves in Fig. 29c.

In both cases of crystallization with a constant thickness of the impoverished zone the surface tension at the phase separation boundary plays an important role. When the thickness of the impoverished zone does not exceed the radius of the critical nucleus, the surface tension which enters into the constant a of Eq. (5.11) plays a role only until the faces of the growing crystal become plane. Up to this moment the crystal grows faster in the direction in which the surface tension is minimum. After faceting, the growth rate, according to Eq. (5.11), is determined by the relationship:

$$v = DV \frac{C - C_{\sim}}{x}.$$

This relationship can be transformed into the Nernst-Andreev equation [122, 123].

If we assume that the solubility limit $C_{\sim}$ depends to a certain extent on the value of the surface tension (one may expect an inverse relationship [142,143]), then the relationship shows that the facets with the highest surface energy will grow at the highest rate after faceting of the crystal is completed. The habit of the crystal after faceting is determined by the Gibbs-Curie-Wolff relationship [144, 145].

When the radius of the curvature remains constant during crystal growth (x = const) the rate of crystal growth depends on the surface tension at all stages of growth, and therefore in this case the crystals have the most complex shapes (skeleton); these shapes are determined essentially by the crystalline structure of the precipitating substance and the structural relationship with the surrounding phase. Let use consider a few examples.

In iron-carbon alloys the primary plates of cementite or graphite grow with a constant radius of curvature during solidification of posteutectic cast irons and dendrites of the austenite of preeutectic cast irons and steels also grow with a constant radius of curvature. As we have said, the main faces of cementite have different surface tensions at the boundaries with the surrounding melt because of the singularities of the crystal lattice. Therefore, the combination of growth rates in mutually perpendicular directions leads to the formation of platelets.

The graphite has a different structure, but the ratio between the surface tensions in different mutually perpendicular directions is about the same as in cementite. Therefore, in gray cast iron the graphite usually has the shape of platelets.

Under the same crystallization conditions austenite crystals grow in the form of dendrites because the crystal lattice has a different structure. In austenite, which has a face-centered cubic lattice, the surface tensions are minimum and equal (at the boundary with the liquid) in the directions perpendicular to the {100} faces.* Therefore, an austenite nucleus having reached a size sufficient to grow with a constant radius of curvature (this size is determined from the condition x = const > r) begins to branch in directions perpendicular to the {100} faces. The branches formed in this manner (first-order axis) grow until their mutual effects stop the supply of structural material necessary for side growth. As soon as this effect ceases, the ends of the branches "sprout" again into four new mutually perpendicular directions, which form second-order axes, etc.

The growth mechanism of spheroidal graphite has certain singularities in cast iron modified with magnesium in sulfurous cast irons, Co–C and Ni–C alloys, and also in cast irons purified by gas injection.

At first glance, the formation of spheroidal graphite in solidifying cast iron contradicts the growth mechanism described earlier because the surface tension of graphite in different crystallographic directions is very different at the boundaries of graphite with the liquid and solid phases. In this case the phenomenon cannot be explained by the adsorptional effect of the modifier because spheroidal graphite forms without any modifier in Co–C and Ni–C alloys.

The main cause of the formation of globular graphite is that the growing inclusions consist of a large number of nuclei in which the basal planes of the atomic lattice of graphite are oriented toward the growing surface.

Thus, at the liquid-graphite crystallization front the surface tension is about the same in all directions (Fig. 30), and thus the growth rate is the same in all directions.

---

* The surface tension of a deformed surface is some average value of the surface tension of a large number of faces adjacent to the surrounding phase in the direction of growth. Thus, the crystallographic directions in which the growth rate is maximum given here are arbitrary.

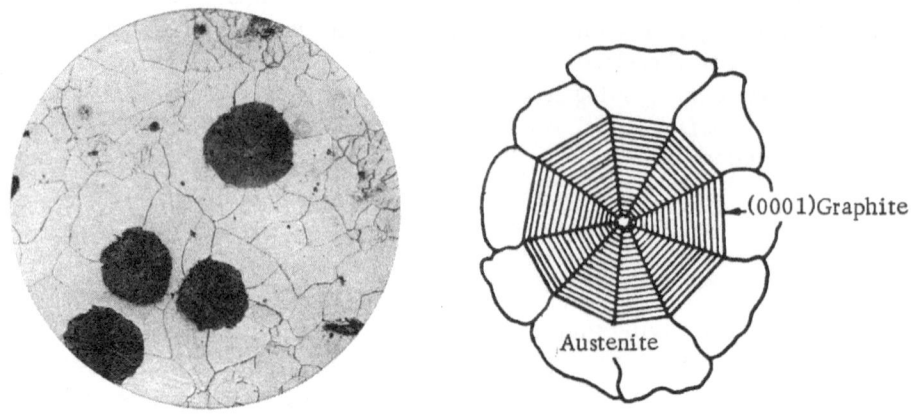

Fig. 30. Illustration of the mechanism of growth of spheroidal graphite.
a) Microstructure; b) diagram of the precipitation of graphite.

We shall show later (see Chapter 8, Section 4) that the nuclei form on the inner surfaces of micropores in the liquid during supercooling of liquid cast iron. These nuclei are the so-called primary graphite precipitates. The inner surfaces of the gas bubbles filled with magnesium vapor or other gases which do not interact with atomic carbon are the most favorable places for the formation of primary graphite. Clearly, oxygen cannot be one of these gases, since it favors the formation of oxide films.

Dendritic crystallization of more viscous liquids and polyhedral crystallization for less viscous liquids are explained by the fact that in viscous liquids the thickness of the impoverished zone is greater. It follows then, that in more viscous liquids the branches of the dendrites must be thinner because the radius of the cylinder corresponding to the maximum growth rate is smaller. It is well known that in practice dendrites crystallized at lower temperatures have thinner branches.

This description of the early stages of crystal growth can also be used to explain diffusionless crystallization (crystallization of unicomponent systems, allotropic transformations, martensitic transformations).

The mechanism of crystallization processes in which diffusion has no significant role can be described as follows. An envelope in which the temperature is different from that of the supercooled mother phase is formed around the growing crystallization center as the result of the evolved heat of crystallization. The conduction of heat away from the crystallization front allows the growth of the crystallization centers. The amount of crystallized substance is determined by the following relationship:

$$q = \frac{Q}{L},$$

where L is the specific heat of crystallization and Q is the amount of heat conducted through the thermal jacket surrounding the growing crystallization center, which is a perfect sphere with inner and outer radii.

By using the analogy between the diffusion equation and the heat conductivity equation we can express the amount of heat conducted from the front of the crystallization center, represented in Fig. 29a by the following relationship:

$$q = 4\pi\chi \int_0^t \frac{(T_1 - T)\, r r_1}{r_1 - r}\, dt, \qquad (5.12)$$

where T_1 is the equilibrium temperature at the crystallization front; T is the temperature at the outer boundary of the sphere, i.e., the temperature of isothermal crystallization; and χ is the coefficient of heat conductivity.

The temperature at the crystallization front can be determined by the well-known relationship

$$T_1 = T_S \left(1 - \frac{2\sigma MV}{Lr}\right), \tag{5.13}$$

where T_S is the melting point of the crystallizing substance; σ is the surface tension; M is the molecular weight; V is the specific volume; and r is the radius of curvature of the separation boundary.

If we substitute Eq. (5.13) into (5.12) and differentiate, we obtain

$$\frac{dq}{dt} = 2\pi\chi \left(T_S - T - \frac{2\sigma MVT_S}{Lr}\right) \frac{rr_1}{r_1 - r}.$$

If $r_1 - r = x$, then the growth rate for a half sphere is

$$v = \frac{dr}{dt} = \chi V \left(\Delta T - \frac{a_1}{r}\right)\left(\frac{1}{x} + \frac{1}{r}\right). \tag{5.14}$$

This equation is similar to (5.9).

Using the same considerations as in the case of isothermal decomposition of supersaturated solutions, one obtains the same result for diffusionless isothermal crystallization.

Thus, for the growth rate of a needle (or platelet) with a constant radius of curvature of the separation boundary r (x being large and constant), we have

$$U = \frac{2\chi V}{L}\left(\frac{\Delta T}{r} - \frac{a_1}{r^2}\right). \tag{5.15}$$

The growth rate of a crystal with constant x which is smaller than the radius of the critical nucleus is

$$v = \frac{\chi V}{Lx}\left(\Delta T - \frac{a_1}{r}\right).$$

Since the heat is conducted at a much higher rate than the diffusion rate of the substance, we may assume that $x \gg r$ in the case of growth of spherical crystallization centers when the degree of supercooling is low.

Then, for this case the growth rate will be

$$v = \frac{\chi V}{L}\left(\frac{\Delta T}{r} - \frac{a_1}{r^2}\right). \tag{5.16}$$

From the same considerations it follows that in the case of diffusionless processes needles, dendrites, platelets, and polyhedrons can occur only when the liquid or the gas is thoroughly mixed or there are intense convection currents. In transformations of the solid state needles and plates are created when the degree of supercooling is high and a thin thermal jacket is formed around the crystal as the result of a high rate of heat conduction.

The relationship between the amount of crystallization heat evolved and the capacity of the medium to conduct this heat away from the crystallization front plays an important role.

When the crystallization heat and the temperature of the medium are low and the heat conductivity of the medium is high, the thermal jacket is thin. If we also take into account the motion of the heat source (crystallization front) in the direction in which the heat is conducted away, then we can assume that the creation of a thermal jacket of constant thickness is quite probable.

Let us also note that "three-dimensional" dendrites almost never form during phase transformations in the solid state because the newly formed phase is oriented in a definite direction with respect to the old phase. In this case the number of directions with low surface tensions (which determine the growth rate in the early stages of transformation) is very limited.

The relationships obtained here concern the growth and the shape of crystals under ideal conditions of isothermal crystallization. In the final analysis, they are the result of the interaction of a large number of factors characterizing the surrounding medium and the growing crystal itself. The interaction between these factors determines the kinetics and the mechanism of crystal growth.

As to the impoverished zone, its existence was proven experimentally in [117, 118], where the authors also showed the possibility of decreasing the thickness of the crystallization enclosure down to molecular size.

Several authors have indicated that the formation of dendrites requires a given degree of supercooling at the crystal-liquid separation boundary [111, 112]. Although the temperature gradient is only one of the many factors affecting the thickness of the crystallization enclosure, it is interesting to find (by relatively simple calculations) the effect of the temperature field on the distribution of concentrations around the growing crystallization center.

The problem in the steady state is reduced to the solution of the equation:

$$\nabla \left(D_1 \nabla C\right) = 0 \quad \text{or} \quad \nabla D \nabla C + D \nabla^2 C = 0.$$

According to the Greene theorem,

$$\int \left(\nabla D \nabla C + D \nabla^2 C\right) dV = \oint D \frac{\partial C}{\partial r}\, dS.$$

In the case of spherical symmetry we have

$$D \frac{\partial C}{\partial r}\, 4\pi r^2 = \text{const}$$

or

$$D \frac{\partial C}{\partial r} = \frac{\text{const}}{r^2}. \tag{5.17}$$

If we neglect the temperature field, this equation gives the ordinary distribution of concentrations when $c_{r=\rho} = C_\rho$ at the boundary of the growing crystal and $C_{r \to \sim} = C_0$ away from the crystal; here, C_ρ is the equilibrium concentration and C_0 is the concentration of the solution. Under these conditions, i.e., in the case where $x \to \sim$, the growth rate is determined by Eq. (5.10). In Eq. (5.17) the diffusion coefficient is highly dependent on temperature:

$$D = D_0 e^{-\frac{Q}{RT}}.$$

Substituting this expression into (5.17), we obtain

$$\frac{\partial C}{\partial r} = \frac{A}{r^2}\, e^{\frac{Q}{RT}},$$

where $A = \text{const}/D_0$.

If the degree of supercooling remains constant during crystal growth then the temperature distribution in the mother phase can be approximated by the following relationship:

$$T = T_0 + \frac{\rho\left(T_\rho - T_0\right)}{r},$$

where

$$e^{\frac{Q}{RT}} = e^{\dfrac{Q}{RT + \dfrac{R\rho\left(T_\rho - T_0\right)}{r}}}$$

We shall transform the exponent in the following way:

$$\frac{Q\left(RT_0 - \dfrac{R\rho\Delta T}{r}\right)}{\left(RT_0 - \dfrac{R\rho\Delta T}{r}\right)\left(RT_0 - \dfrac{R\rho\Delta T}{r}\right)} \approx \frac{QRT_0 - \dfrac{R\rho\Delta T Q}{r}}{R^2 T_0^2},$$

then

$$e^{\frac{Q}{RT}} = \text{const} \cdot e^{-\frac{\rho \Delta Q}{RT^2 r}}$$

By introducing the notation:

$$\frac{\rho \Delta T Q}{RT_0} = K_\rho,$$

we obtain

$$C = A_1 \int \frac{e^{-\frac{K_\rho}{r}}}{r^2} + B, \tag{5.18}$$

where B is the integration constant,

$$A_1 = A \cdot \text{const.}$$

Equation (5.18) can be transformed into

$$C = A_1 \int e^{-\frac{K_\rho}{r}} d\left(-\frac{K_\rho}{r}\right) + B,$$

and then

$$C = A\, e^{-\frac{K_\rho}{r}} + B.$$

Let us examine this relationship. The first derivative

$$\frac{dC}{dr} = A\, e^{-\frac{K_\rho}{r}} \cdot \frac{K_\rho}{r^2}$$

indicates that C increases with increasing r, and consequently C(r) has no extremum. The second derivative

$$C'' = A \left[\left(\frac{K_\rho}{r^2}\right)^2 e^{-\frac{K_\rho}{r}} - \left(\frac{2K_\rho}{r^3}\right) e^{-\frac{K_\rho}{r}} \right]$$

becomes zero for a given value of r. Thus, C(r) has an inflection point (Fig. 31). Let us determine the x coordinate of the inflection point. To do this we make C'' = 0 and determine r_0:

$$r_0 = \frac{K_\rho}{r} = \frac{\Delta T Q}{2RT_0^2},$$

and x is given by

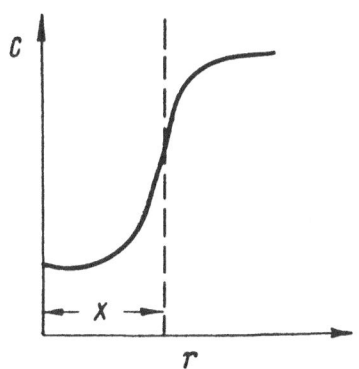

Fig. 31. Distribution of concentration at the crystallization front.

$$x = r_0 - \rho = \rho \left(\frac{\Delta T Q}{2RT_0^2} - 1 \right). \tag{5.19}$$

This value is the thickness of the crystallization enclosure; it is zero when K > 1. As an example, let us investigate the precipitation of cementite from austenite in a posteutectic steel at temperatures below the SE line on the Fe−C phase diagram. For a supercooling $\Delta T = 150°C$ at Q = 32000 cal/deg and $T_0 = 1000°K$ (the steel contains ~1.2% C), x = 0.2ρ. The value of x increases with the degree of supercooling.

For a maximum growth rate ρ must be constant during the formation of a crystallization enclosure with a constant thickness. From the conditions of maximum v in (5.9) we find $r_m = \rho_m$. Substituting this result into Eq. (5.19), we obtain the thickness of the zone x, namely $x = r^*(Q\Delta T/RT^2 - 1)$. It is interesting that in Eq.(5.17) the value of ΔT is determined by the amount of crystallization heat evolved. Therefore, only a rather high degree of supercooling can lead to the formation of a crystallization enclosure. In fact, cementite grows in the form of needles in posteutectoid steel only when the degree of supercooling is very high. In these calculations we took only the temperature field into account. The dependence of the diffusion coefficient on the concentration complicates the conditions under which the crystallization enclosure forms. It is also important to take into account convection and thermal diffusion of impurities [146]. All these problems remain to be solved.

Calculations showed that the thickness of the thermal jacket or the impoverished zone may be used as a criterion of the type of crystal growth.

Thus, if the thickness of the impoverished zone is constant and smaller than the radius of the crystal the globular nucleus will in the end be transformed into a polyhedron. If the thickness of the zone is constant and larger than the radius of the nucleus then one obtains needles, platelets, dendrites, depending on the structure, i.e., the curvature of the crystallization front remains constant. When the thickness of the impoverished zone increases then the radius of curvature of the crystallization front also increases.

The mechanism by which some modifiers affect the type of crystal growth can be explained on the basis of the considerations given here. In fact, a small amount of low-melting-point impurity can change the thickness of the impoverished zone or of the thermal jacket as the result of thermal diffusion and, as the consequence, change the shape of the growing crystal.

4. Crystal Growth during Phase Transformation

The growth rate of crystals and the mechanism of the formation of their shape described above refer to ideal crystallization conditions. The growth of nuclei during phase transformation of ordinary alloys is complicated because in practice the systems are not ideal. Concentration current, the mutual effect of growing crystals, liquation, complex temperature fields, soluble and insoluble impurities — all these factors change the rate of crystal growth by several orders of magnitude.

Even under relatively "pure" laboratory conditions the use of the equations derived earlier to determine the rate of crystal growth during phase transformation is limited because the principal singularity of crystal growth during phase transformation is the change in the composition of the crystallization centers. This change in the composition is subject to complex laws which, in turn, determine the change of the thickness of the crystallization enclosure.

There are very few quantitative studies of these laws. As an example, we can cite the comparison between the theoretical and experimental data on the growth of ferrite grains. This is of particular interest because the rather precise calculation (nonsteady state) of the rate of grain growth described in [126] was checked experimentally. As we have noted before, supercooling plays an important role in the precipitation of crystals from supersaturated solutions. The degree of supercooling affects not only the degree of supersaturation but also has a great effect on the parameters of the crystallization enclosure.

Theoretical results were checked in [147] by studying the kinetics of the precipitation of ferrite from supersaturated austenite. In this study the author used carbon steel rods containing 0.15%, 0.3%, and 0.45% C. Flat crescent-shaped samples with a radius of 6 mm and a thickness of 3 mm were cut out of these rods.

The samples were heated in lead baths. First the samples were immersed for 20 min in a bath at a temperature above the A_3 point (900°C). Then the samples were transferred to a bath at a temperature of 736°C, where ferrite crystallization centers began to grow. The temperature fluctuation did not exceed ±2°.

The samples were then quenched in water. A layer 2 mm thick was removed from the end of the samples, the samples were etched, and the average thickness of the ferrite layer was measured. The growth rate of ferrite grains can be calculated by Eq. (5.9).

TABLE 3

Time at the given temp., min	Growth rate, $v \cdot 10^6$ cm/sec	
	theoretical	experimental
Steel with 0.15% C		
4 sec	8.6	24
0.6	3.2	16
0.3	3.8	12.5
0.5	2.5	11.5
1	2.2	11
Steel with 0.3% C		
2	0.7	3.3
3	0.6	1.5
5	0.4	0.5
6	0.35	0.3
7	0.3	0.1
Steel with 0.45% C		
2	1.5	2.5
2.5	1.1	1.7
3	0.9	1.4
4	0.6	0.75
5	0.3	0.4

TABLE 4

Time at the given temp., min	Growth rate, $v \cdot 10^6$ cm/sec	
	theoretical	experimental
5	$6.4 \cdot 10^{-6}$	$80 \cdot 10^{-6}$
10	$5 \cdot 10^{-6}$	$65 \cdot 10^{-6}$
20	$2.5 \cdot 10^{-6}$	$35 \cdot 10^{-6}$
30	$2 \cdot 10^{-6}$	$25.5 \cdot 10^{-6}$
40	$0.71 \cdot 10^{-6}$	$18 \cdot 10^{-6}$
60	$0.64 \cdot 10^{-6}$	$17 \cdot 10^{-6}$

Since the author investigated only the later stages of growth, the following equation was used after integrating and neglecting small terms:

$$v = \sqrt{\frac{DV(C - C_\sim)}{2t}} . \qquad (5.20)$$

The theoretical and experimental data are shown in Table 3. Comparison of these data indicates that the agreement between the data improves with increasing concentrations of carbon and decreasing degrees of supercooling (since the temperature of isothermal treatment is the same, the degree of supercooling decreases with increasing concentrations of carbon in the steel).

Let us compare the theoretical data with the experimental data obtained in [146]. In this investigation the samples were supercooled to a much greater degree, and therefore one would expect a much poorer agreement of the results. The experimental data obtained in [146] and the theoretical data obtained by using Eq. (5.20) are shown in Table 4. It can be seen that the difference between the experimental and theoretical data exceeds one order of magnitude. The same results are obtained when the calculations are made by the equation proposed in [126].

The ferrite-austenite transformation is characteristic of the decomposition of supersaturated solid solutions. The principal transformation in the decomposition of solid solutions is the polymorphic transformation of γ-iron into α-iron. The mechanism of this transformation consists in the rebuildinf of the face-centered cubic lattice of the γ-phase into a body-centered lattice of the α-phase. The rate of this transformation is very high because the transformation results from the motion of atoms over distances not exceeding the lattice constant. However, the presence of carbon has a great effect on the kinetics of this process. The solubility of carbon in ferrite is very low ($\sim 0.03\%$); therefore, when the degree of supercooling is low, ferrite grains do not capture carbon during their growth, carbon accumulates at the crystallization front, and the growth rate of ferrite grains can be determined by their relationships relative to the decomposition of supersaturated solid solutions. The agreement between the theoretical and experimental data must decrease with increasing degrees of supercooling because when the degree of supercooling is high the rebuilding of the lattice is the principal process and the ferrite grains capture carbon atoms during their growth. It is quite probable that ferrite grains are surrounded with a crystallization enclosure whose thickness decreases with increasing degrees of supercooling. This hypothesis is confirmed by the fact that ferrite grains become needlelike with increasing degrees of supercooling.

Experimental results indicate that the diffusional theory of the decomposition of supersaturated solid solutions in which the thickness of the crystallization enclosure is not accounted for can be applied to the austenite-ferrite transformation only when the degree of supercooling is low and when the carbon concentration is relatively high.

In the case of high degrees of supercooling, the most important role is played by the processes of rebuilding of the γ-lattice into the α-lattice. But even in this case, the presence of carbon modifies the degree of supercooling, the value of the surface tension, and many other magnitudes which determine the kinetics of the process, and the ferrite grains grow according to the laws of martensitic transformation.

We shall not go into the details of the mechanism and the kinetics of the martensitic transformation, but study only those relationships which directly concern the conclusions given in this chapter.

1. Martensite crystals have the shape of needles (or plates) [148].

2. In a steel containing 29.5% Ni, 0.027% C, 0.136% Mn, and 0.094% Si the growth rate of martensite needles is of the order of ~10^5 cm/sec. The size of these crystals remains practically constant between 20 and 195°C [149].

3. In some cases the growth rate of the needles depends on the variation of the concentration.

According to Kurdyumov's theory, transformations of the martensitic type obey the ordinary laws of phase transformation. This transformation begins with the formation of crystallization centers and their subsequent growth.

There is no complete agreement on the mechanism and the kinetics of the formation of martensite crystallization centers, which are not yet completely known. As to the growth rate, the singularities noted earlier are commonly accepted.

The concepts of the mechanism and kinetics of crystal growth developed in this chapter confirm the validity of that part of the Kurdyumov theory of martensitic transformation which concerns crystal growth.

In fact, when diffusion does not intervene in the growth of martensite needles their growth rate is determined by the rate of conduction of heat away from the crystallization front. The growth rate of a needlelike crystal, i.e., a crystal which grows with a constant radius of curvature, is described by the equation:

$$\frac{dh}{dt} = \frac{2\chi}{L}\left(\Delta T - \frac{a_1}{r}\right)\left(\frac{1}{x} + \frac{1}{r}\right).$$

When $x \gg r$, which is the most probable situation in solid systems, the growth rate is described by the equation:

$$v = \frac{dh}{dt} = \frac{2\chi}{L}\left(\frac{\Delta T}{r} - \frac{a}{r^2}\right).$$

If in this relationship we replace r with r* n, where r* is the radius of the critical nucleus and n is the number of times the radius of curvature of the crystallization front is greater than the radius of the critical nucleus, we have

$$U = \frac{2\chi\Delta T^2}{La_1}\left(\frac{n-1}{n^2}\right),$$

since r* = a_1/Δ. Since we assumed that the thickness of the thermal jacket is much larger than r, it is clear that the growth rate is maximum when n = 2.

Then

$$v = \frac{1}{2} \cdot \frac{\chi\Delta T^2}{La},$$

and after substitution of

$$a = \frac{2\sigma T}{L},$$

where L is the crystallization heat per unit volume, we have

$$U = \frac{\chi\Delta T^2}{2\sigma T_S},$$

where σ is the surface tension at the martensite-austenite boundary.

Thus, the growth rate of martensite crystals depends on the coefficient of heat conductivity, on the degree of supercooling, on the martensite-austenite equilibrium temperature, and also, as one would expect, on the surface tension.

For all alloys in which the martensitic transformation occurs there is a temperature T_S at which the free energy of the initial solid solution is equal to the free energy of the martensite phase (see Chapter 6).

For most alloys it is very difficult to determine T_S experimentally because the martensite undergoes diffusional decomposition during heating. If T_S happens to be a low temperature then equilibrium conditions cannot be reached in most cases and the alloy must be superheated to some temperature A_S (determined for the Fe–C alloy in [149]) for the alloy to return to its initial state. The precision in the determination of T_S depends on the difference between the temperatures M_S (martensitic transformation temperature) and A_S.

The growth rate of martensite needles was measured in Fe–Ni alloys and the value of T_S for these alloys was calculated in [150].

The value of T_S is close to 450°K for alloys containing about 29.5% Ni and $\Delta T \sim 200$°K at –20°C. The heat conductivity is $\chi = 2 \cdot 10^{-2}$ kcal/cm · sec · deg.

Using these values and the value of the growth rate of martensite needles determined in [149], we calculated the surface tension at the martensite-austenite boundary, which turned out to be equal to:

$$\sigma = \frac{2 \cdot 10^{-2} \cdot 4 \cdot 10^4}{4 \cdot 10^5 \cdot 450} \approx 180 \quad \text{dynes/cm} .$$

This value agrees with the most probable value given in [151]. Although this calculation is approximate, it does show that the extremely high growth rate of martensite crystals does not contradict the fundamental laws of crystallization and that it can be obtained from the ordinary equation in the case of diffusionless transformations [152].

From Eq. (5.13) it becomes clear why the change in the growth rate with increasing degrees of supercooling was not observed in the experiments described in [124].

In fact, when the degrees of supercooling is increased to 450°K (i.e., the transformation occurs almost at absolute zero) the value of ΔT increases from $4 \cdot 10^4$ to $2 \cdot 10^5$°K. At the same time, the surface tension at the austenite-martensite boundary increases. According to [267] and [268], the surface tension can reach 1000 ergs per cm^2 without impairing the coherence (as defined by Kurdyumov). It is clear that the loss of coherence leads to a sharp increase in the surface tension and the martensite crystal stops growing. Apparently, the main reason the martensite crystal stops growing is that elastic stress changes the position of the T_S point.

The mechanism of crystal growth described in this chapter refers to the early stages of the development of crystallization centers. This description is a generalization of a large number of unsystematized experimental results. Some of the conclusions given here are qualitative and even hypothetical, since their confirmation requires stricter calculations and a greater number of experimental results. However, this description of the early stages of crystal growth is useful, since at the present time there is no precise theory.

CHAPTER 6

STRUCTURE FORMATION IN ALLOYS

1. Intracrystalline Liquation

In the preceding chapter we examined the mechanism of the growth of crystals under conditions where the composition of the supersaturated solution and that of the crystallization centers remain constant during crystallization. However, in most alloys phase transformations are accompanied by changes in the composition of the initial and precipitated phases.

The formation of the structure of an alloy is complicated by intracrystalline liquation, a phenomenon which has been known for a long time. The degree of intracrystalline liquation* depends on many factors, among which are the shape of the phase diagram, the cooling rate of the alloy, the diffusion rate in different phases, etc.

There are many experimental data on the effect of these factors on the degree of intracrystalline (dendritic) liquation and all these data are related by semiempirical equations. A systematic description of these problems was given in [153]. Unfortunately, the physical foundations of intracrystalline liquation are not usually described. This is understandable, since the modern crystallization theory does not yet make it possible to calculate the growth rate of crystals under complex practical conditions.

For a complete description of the formation of the structure of alloys we shall study the process of intracrystalline liquation qualitatively by using hypothetical free energy curves.

Experiments have shown that the degree of intracrystalline liquation, other conditions being equal, depends on the cooling rate in the manner shown in Fig. 32a. The maximum on the curve, corresponding to the highest degree of liquation, occurs when diffusion in the precipitated phase is very low, while it is high in the initial phase. Let us consider the case of crystallization in a system with unlimited solubility in the solid and liquid states. The solid phase II precipitates from the supercooled liquid phase I with a concentration C_0. When the cooling rate is low, the nuclei of phase II have time to precipitate during the first stage of cooling, and consequently their composition corresponds to C_1 (see Fig. 32b), since at this degree of supercooling the tangents $\Delta \mu_A = \Delta \mu_B$ are parallel only for this composition. During the growth of the nucleus at constant temperature T_1 the composition of the nucleus changes in the direction indicated by the arrow at II and in the liquid phase by the arrow at I, provided the condition $\Delta \mu_A = \Delta \mu_B$ is satisfied at every given moment.

This clear interrelationship between the concentrations of the liquid and solid phases exists only in a very narrow region next to the crystallization front (quasiequilibrium state). Away from the crystallization front, the concentrations in the liquid and solid phases are determined by the diffusion rate of the substance, which equalizes the concentration in the whole volume of the liquid and solid phases. The time in which the concentration is equalized by diffusion is determined by the rate of cooling of the alloy to the temperature at which the mobility of atoms becomes very low.

Ordinarily, one considers the three following situations: 1) low cooling rate, when diffusion in the solid as well as in the liquid phase has time to equalize the concentration so that at the end of crystallization the concentration of the solid phase is the same as that of the initial liquid; 2) high cooling rate, where at some point diffusion in the solid phase does not have time to equalize the concentration, while the concentration in the liquid phase is equalized. Then, the composition of the solid phase is not uniform, and as the result each newly crystallized layer is in equilibrium with the liquid, whose composition is continually enriched in the

---

* The degree of intracrystalline liquation is the ratio between the concentrations at the inner and outer parts of the crystal.

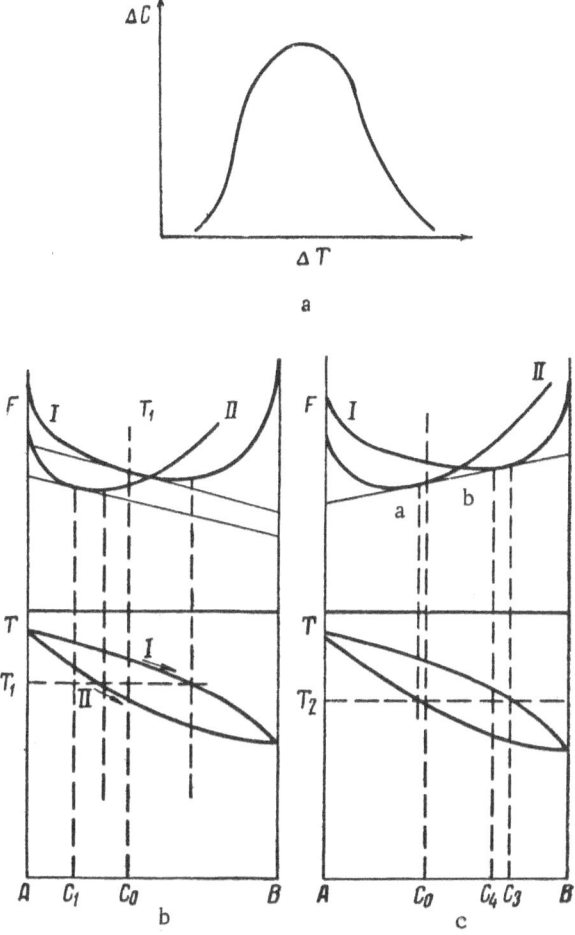

Fig. 32. Effect of the degree of supercooling.

B component. The degree of liquation depends on the initial degree of supercooling. The higher the temperature at which the nuclei are formed, the greater the microscopic liquation. The transformation at high cooling rates is the most interesting, since it has not been extensively studied; 3) when the cooling rate is very high, the nuclei of the solid phase occur when the degree of supercooling is very high, but still above the solidus curve. Therefore, the distance between curves I and II is large (Fig. 32c), and $\Delta \mu_A = \Delta \mu_B$ is much greater.

This will lead first of all to a higher rate of formation of dendrites and, since the thickness of the crystallization enclosure increases with the degree of supercooling, the dendrites will be very thin in the early stages of growth. Since the growth rate of the dendrite axis is very high, atoms of the B component will accumulate on the outer side of the enclosure up to the concentration C_3. Further growth and thickening of dendrite branches can occur only when the equilibrium conditions, determined by tangent ab, are satisfied. The equilibrium concentration of dendrites will then be C_0, while the concentration of the liquid at the outer boundary of the enclosure will be C_3. At this point, the kinetics of growth and of the thickening of dendrites will be determined by the decrease of the concentration gradient from C_3 to the rest of the liquid. When the composition of the surface layer of dendrites is the same as that of the main part of the liquid – aside from the narrow region surrounding the crystallization enclosure, where the concentration is C_3 (Fig. 33, zone Z) – the concentration of the solid phase cannot increase any longer because the number of B atoms in the zone (with a concentration C_3) cannot increase. From this moment onward, the B atoms migrate into the rest of the liquid, with a concentration C_0. However, as soon as the concentration of the liquid in the Z zone decreases (let us assume to C_4 – see Fig. 32c)

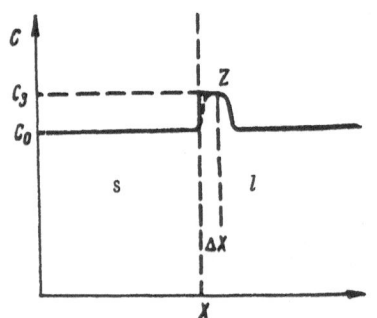

Fig. 33. Distribution of concentrations at the front of a crystal growing at a high degree of supercooling.

the liquid immediately becomes unstable and a solid phase with a concentration lower than C_0 but very close to it begins to precipitate. At the same time, the decrease in the number of B atoms in zone Z is compensated because the solidifying part of the liquid is slightly enriched in the A component.*

Thus, dynamic equilibrium is established. This equilibrium consists in the fact that the B atoms move from the Z zone into the liquid with a concentration C_0 and simultaneously promote growth of the crystal. The more enriched the zone, the higher the rate of this process. On further cooling, zone Z itself solidifies and the concentration is increased by the same amount as it is decreased at the initial moment of crystallization.

The experimental data apparently confirm this mechanism of growth, since it was indicated in [153] that a decrease of diffusion in the liquid phase leads to greater concentration gradients of the impurity in the initial and final layers of the solid phase, while the concentration of the impurity is relatively uniform in the middle part of the solid phase.

All this shows that further development of the theory requires study of the distribution of the temperature and concentration at the crystallization front.

Since a large number of investigations and textbooks [154, 155] concern the formation of structure in alloys at relatively low cooling rates (~1000°C/sec), we shall examine crystallization of alloys at high cooling rates (≫ 1000°C/sec).

The study of the formation of the structure of alloys at high cooling rates has not only a theoretical value but also great practical interest, since the crystallization of metals in processes such as welding, chill casting, production of sheet cast iron, continuous pouring, etc., occurs at very high cooling rates.

2. Some Concepts of the Formation of the Structure of Alloys at High Degrees of Supercooling

Many attempts have been made to explain the mechanism of the formation of the structure of alloys solidified as the result of a high degree of supercooling. Most of these attempts consist of qualitative analyses of the kinetics of the formation and growth of crystals by extending the lines on the phase diagram into subcritical regions to determine the compositions of the crystallizing phases. Figure 34 shows a sample of such a phase diagram. If, for example, we examine the crystallization of a supercooled liquid at temperature T_1, the liquid with a concentration lower than C_1 and greater than C_2 must be transformed into a solid solution of the same concentration. A quasi-eutectic must form between C_1 and C_2. The dashed line indicates the phase equilibrium in the subcritical region, according to which a supersaturated solid solution must precipitate from a highly supersaturated liquid. The extension of the liquidus line into the subcritical region indicates the supersaturation boundary of the liquid in equilibrium with one phase when the second phase is not yet precipitated. In [156, 157] it was shown that extension of the liquidus line into the subcritical region is valid. However, the validity of the extension of the solidus line into the subcritical region has not been demonstrated. The maximum concentration of the α-phase is C_m and, according to the concept based on the curves of free energy, this concentration is the equilibrium concentration of three phases (Fig. 35a). When the crystallization temperature is lowered to T_2 the solubility of the α-phase decreases, and therefore the concentration in the solid solution also decreases, although the liquid solution is supersaturated with the B component up to concentration C_2. The degree of supersaturation of liquid solutions decreases because of the decrease in the amount of the precipitated α-phase.

---

* If the degree of supercooling is sufficiently high, new crystals can nucleate and grow in zone Z, forming a layer of small-grain metal similar to the outer crust of frozen crystals in an ingot. After this layer is formed, dendrites begin to grow, forming a new layer of columnar crystals.

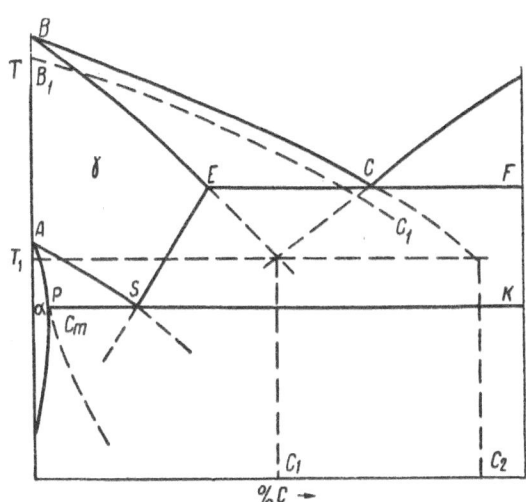

Fig. 34. Phase diagrams with extensions of the equilibrium lines into the supercritical region; B_1C_1) metastability limit.

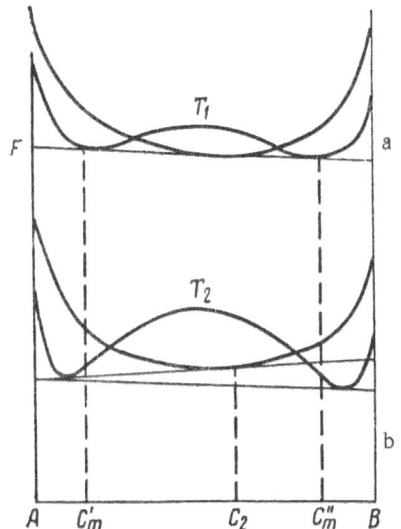

Fig. 35. Limit solubility of the primary solid solution.

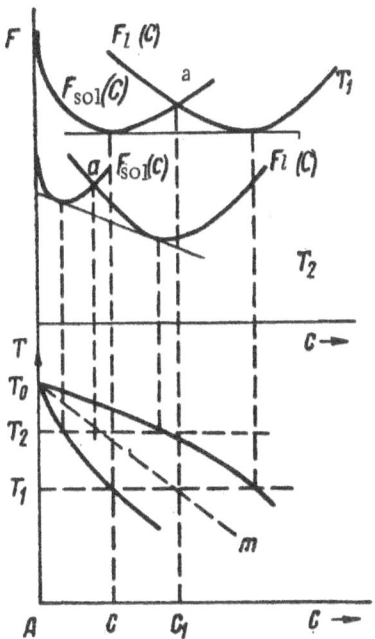

Fig. 36. Line of equal free energies $T_0 m$.

The attempt to substantiate the possibility of increasing the concentration of the primary solid solution above C_m was formulated in the so-called theory of diffusionless crystallization.*

As early as 1913, Hasselblatt [158] used the main principle of this theory to explain some singularities of the crystallization process in systems with unlimited solubility in the solid and liquid states.

In [159, 161] the theory was extended to systems of the eutectic type. However, at present there are many experimental data which do not fit into the theory of diffusionless crystallization, and therefore some aspects of its applicability are doubtful. Consequently, it seems of interest to discuss the foundations of this theory in more detail. The basic concept of the theory is illustrated in Fig. 36.

The $F(C)$ curve represents the dependence of the free energy of a liquid binary alloy of a "cigar" type system on the concentration at some temperature T. The $F_{sol}(C)$ curve corresponds to the solid state. The intersection point a determines the equality of the free energies of the liquid and solid phases. $F_{sol}(C) < F_{liq}(C)$ for all concentration of the alloy less than C_1, and therefore the solid state is thermodynamically more advantageous. The geometric locus of the intersection points represents the line of equal free energies $T_0 m$. According to the theory, the diffusionless transformation of the liquid into the solid phase occurs when the liquid is supercooled below $T_0 m$.

---

* In what follows C_m indicates the maximum solubility of the α-phase at the eutectic temperature in a stable phase diagram. Primary solutions with concentrations above C_m will be called greatly supersaturated solid solutions.

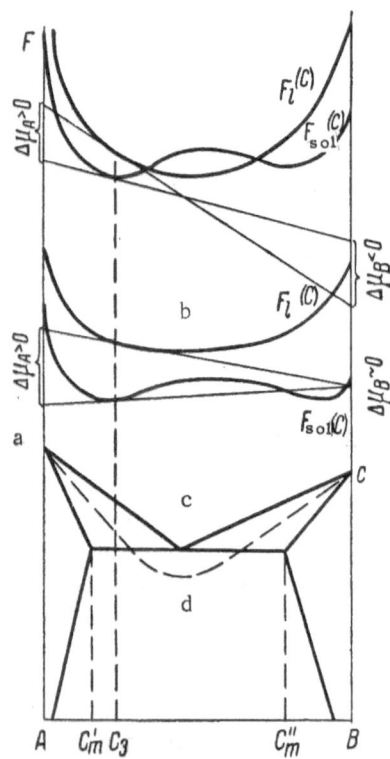

Fig. 37. Line of equal free energies
of a eutectic system [161].

The author of [160] gives calculations on the basis of which this assumption can also be extended to systems of the eutectic type. In this case $F_{sol}(C)$ has two peaks (Fig. 37a) and the line of equal free energies abc is shown in Fig. 37c.

Supercooling of the melt below the abc line leads to diffusionless crystallization regardless of the composition.

Let us note that the line of equal free energies at concentrations between C'_m and C''_m are possible because, according to the theory of the construction of phase diagrams, one can always reach such a degree of supercooling that all the points of the $F_{liq}(C)$ curve will be above the $F_{sol}(C)$ curve (Fig. 37b). It is assumed that the mixing energy, which enters in the equation, is independent of temperature.

Let us state the important conclusions following from the theory of diffusionless crystallization:

1. Alloys with a concentration between C and C_1 solidify completely at a temperature T_1, i.e., above the solidus line (Fig. 36).

2. Primary solid solutions with concentrations exceeding C'_m and C''_m can occur in systems with limited solubility in the solid state (Fig. 37c).

3. Increase of supercooling (or the cooling rate) favors the formation of highly supersaturated solid solutions up to the point where the liquid of eutectic composition is transformed into a solid solution of the same concentration.

From the viewpoint of the thermodynamics of irreversible processes, this means that when a system is supercooled below the abc line (Fig. 37) the liquid phase can be transformed into the solid phase of the same concentration even when $\Delta \mu_A > 0$ and $\Delta \mu_B < 0$, i.e., when the forces acting on the atoms of the B component make them leave the solid solution (for example, in the case of concentration C_3).

Let us now discuss the experimental facts.

3. Experimental Results on Structure Formation in Alloys Crystallized from Highly Supercooled Solutions

The following experimental data, known for rather a long time, are usually cited as a confirmation of the theory of diffusionless crystallization.

1. Intracrystalline liquation and the amount of eutectic in the structure of some alloys decrease as the rate of cooling of the melt increases [161-163].

2. Highly supersaturated primary solid solutions are formed in some electrodeposited alloys [164, 165], in metals prepared by evaporation in vacuum [166], and in cases of extremely high cooling rates [167, 168].

3. In [169] a liquid Hg—Na alloy was transformed into the solid state above the temperature of the solidus line of a stable phase diagram.

4. Transformations of the martensitic type occur without diffusion.

And finally, there recently appeared a paper [170] describing the diffusionless transformation of the liquid phase into a solid solution of the same composition in the Cu—Ag system.

Let us examine these data in more detail.

The first group of facts cannot confirm the principal consequence of the theory described in the preceding section because in all investigations in which crystallization occurred above the temperature of the solidus line there was always a certain amount of liquid phase.

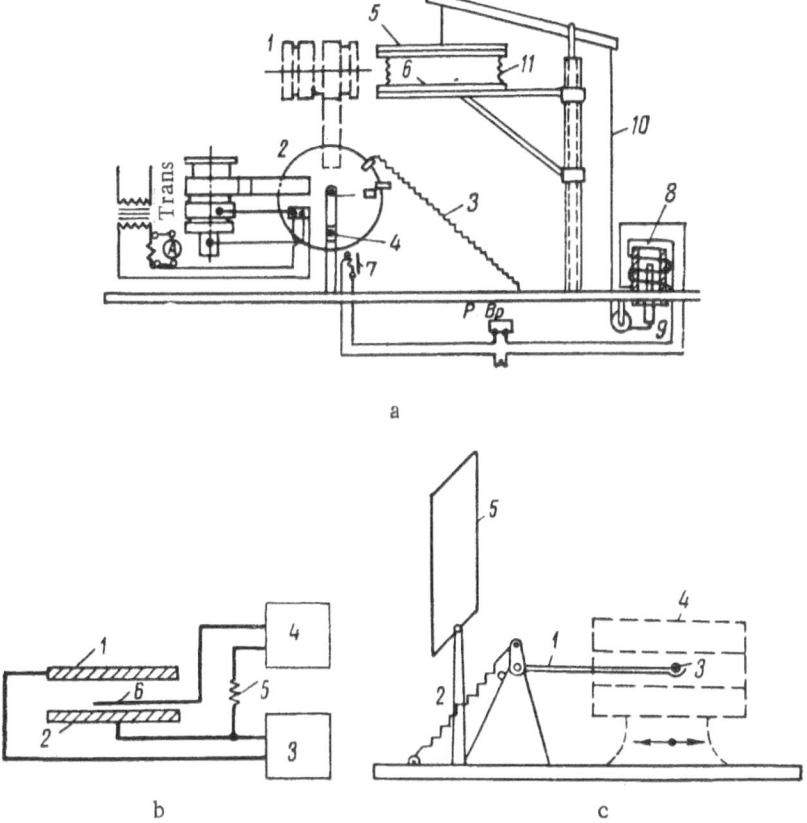

Fig. 38. Apparatus for very rapid crystallization of melts. a) Diagram of the apparatus; b) microsecond stop watch; c) catapult.

Also, it is almost impossible to observe intracrystalline liquation when it is concentrated in narrow regions of the initial dendrite threads or in the Z zone. If such an alloy solidifies at a temperature far below the solidus line then the transformation can occur without diffusion. The study of the formation of the structure of layered systems revealed the following facts.

In the case of simultaneous electrodeposition of Ag and Pb the precipitated solid solution based on Ag has a Pb concentration as high as 9%, while the limit solubility under equilibrium conditions does not exceed 1.5%. We shall limit our examination of electrodeposition to only one example because the mechanism of the formation of alloys in this case is very different from the ordinary mechanism in that here one must take into account an additional factor – overvoltage. This complicates the thermodynamic approach to the crystallization process as if the crystallization occurred under high pressure. Clearly, the phase diagram in this case will not be the same as the stable phase diagram. However, this anomalous increase in the concentration of the primary solid solution above C_m has been observed only in Ag–Pb alloys.

It should be noted that these metals have the same lattice, very similar atomic diameters, but different valences. It is important that an intermediate phase was found in this alloy [171], since the existence of this phase is a decisive factor in the formation of highly supersaturated primary solutions.

In [166] a highly supersaturated solid solution based on Cu was obtained in a Cu–Al system by the Vekshinskii method. This system also contained intermediate phases.

Let us now discuss the data on crystallization from melts. In [167] a vacuum chill mold was used to obtain high cooling rates. This chill mold is a massive copper vessel cooled with water or liquid air. The vessel terminates in a long tube closed with aluminum foil. When the mold is evacuated the tube is lowered into the

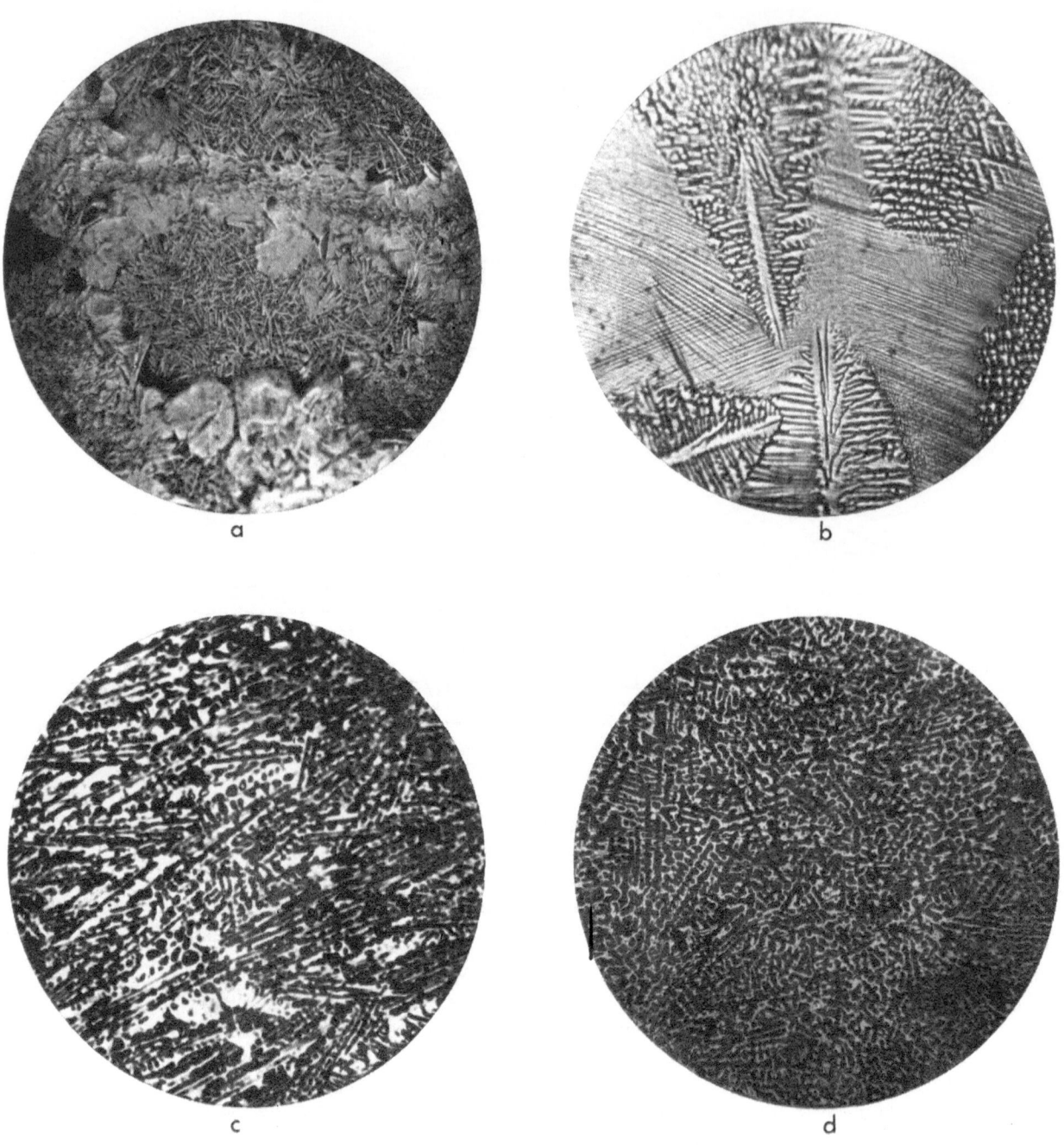

Fig. 39. Structures of alloys crystallized at high cooling rates. a) Surface of the Fe—C alloy, × 500; b) silumin crystallized by catapult, ×500; c) edge of the film of a pre-eutectic cast iron, ×1400; d) separate crystallization of pre-eutectic cast iron, ×1400.

crucible containing the molten alloy. The aluminum foil melts and the liquid metal is drawn into the mold and cools at a high rate on the inner walls. The cooling rate is determined by a thermocouple connected to a recording oscillograph. The cooling rate was of the order of $\sim 5 \cdot 10^4$ deg/sec. Pure metals were supercooled to an extremely high degree by this method. The compositions of the primary solutions of the solidified alloys were determined by the x-ray method. Primary solid solutions with a concentration exceeding C_m have been found. Thus, in Al—Ti the maximum solubility of Ti in Al is $C_m = 0.15\%$ (by weight). After solidification in a vacuum chill mold it increased to 0.32%.

In the Al–V alloy the solubility limit of the α-solid solution based on Al increased from 0.37 to 1.11% V, and in the Al–Cr alloy the solubility limit was increased from 0.7 to 5.5%. The highest degree of supersaturation was reached in the Al–Mn alloy. In this case the solubility limit of the solid solution based on Al increased from 1.4 to 9.2%. In Pb–Na alloys it increased from 1.2 to 3.3%, in Pb–Te alloys from 0.1 to 0.2%, and in Pb–Cu alloys from 0.1 to 1.18%. In Cu–Cr alloys the solubility of Cu increased from 0.7 to 1.45%.

All these alloys, which are of the eutectic or peritectic type, form intermediate phases.

The authors of this investigation did not succeed, however, in obtaining supersaturation above the maximum (C_m) in the Al–Si system, which was described in [159, 160] as a system in which diffusionless crystallization is possible in the case of a highly supersaturated primary solid solution.

In [172] the authors found a strongly supersaturated solid solution based on Zn in Zn–Fe and Zn–Co alloys quenched from the liquid state.

D. S. Kamenetskaya, studying the supercooling of Hg–Na alloys, showed that the solubility of Hg in Na is abnormally high. In this unique experiment of its type it was shown that the supercooled liquid solidifies completely above the solidus temperature. After cooling and reheating it was found that the supersaturated solution can easily be heated above the solidus temperature and that it decomposes only after prolonged heating at this temperature.

The structure and composition of the phase components of films of steel, cast iron, and Al–Si, Al–Zn, Pb–Sn, Bi–Sn, Ni–C, and Co–C alloys solidified at a very high cooling rate were investigated in [173-177].

In this investigation the high cooling rate was reached by using the apparatus shown in Fig. 38. A small Tamman furnace with a magnesite crucible to melt the metal 1 is fixed to a disc 2. The furnace is heated with a current from a welding transformer. When the current intensity is 200 A the furnace temperature rises to 1800°C in 1-2 min. When the spring 3 is released from the stop 4 the disc rotates the furnace from the vertical to the horizontal position (this position is shown in the figure by dotted lines). The metal thrown out of the furnace flies between two copper plates 5 and 6. At the moment the furnace is placed in the horizontal position the relay 7 closes the circuit of the coil of the electromagnet 8 and the core 9 is drawn into the coil and releases the lever 10. At the same time, the upper plate 5, activated by the spring 11, hits the other plate 6 and flattens the molten metal into a thin film.

Chemical analysis of these films showed that the composition of the alloy remains practically unchanged during melting and the transformation into a thin film of solidified metal. The surface of the film is not oxidized during solidification and its structure can be studied under a microscope without preliminary grinding and polishing.

The most characteristic structures of the surface of the film are shown in Fig. 39a and b. The thinnest film obtained by this method is 0.02 mm thick.

The high cooling rate results from the conduction of the heat evolved by the liquid film through the surface of contact with the copper plates, which have a high conductivity and a high heat capacity. The amount of material per unit surface of contact is so small that the heat evolved is insufficient to heat the copper plate to any significant extent.

The cooling rate of the inner layers of the film depends on the coefficient of heat conductivity of the alloy and is somewhat lower than the cooling rate of the surface, but is still very high. As the result, the inner layers of the metal solidify when the degree of supersaturation is lower. A qualitative study of the formation of the structure of the alloy as a function of the degree of supercooling was made by micrographic examination of sections of the films.

The order of magnitude of the cooling rate was determined indirectly by measuring the rate of motion of the upper plate with a stop watch graduated in microseconds. The setup for measuring the rate of motion of the plate is shown in Fig. 38b. The plates, 1 and 2, are insulated from each other and are connected to a contact pulse counter 3. One of the contacts of the counter is connected to a pulse generator 4 through a resistance 5 and the second contact is connected to an intermediate contact made of copper foil. The intermediate

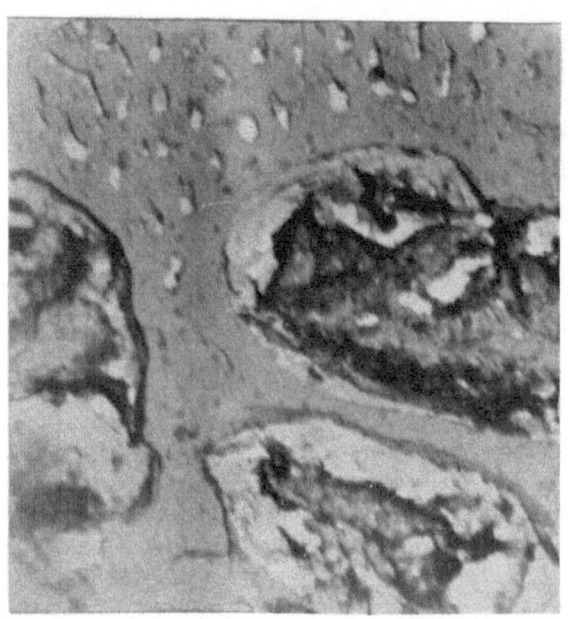

Fig. 40. Electron micrograph of pre-eutectic cast iron solidified at a high rate. ×6000.

contact is placed at different distances from the lower plate. When the upper plate moves it first touches the intermediate contact 6. This connects the generator 4 with the pulse counter 3 and the pulses begin to be counted. At the next moment of the motion of plate 1 the intermediate contact 6 and the plate 2 are connected, the pulse counter is shorted, and the pulse count ceases. The time in which the plate 1 moves from the intermediate contact to the lower plate 2 is calculated from the known frequency of the signals generated and the number of pulses counted.

The rate of motion of the plate was 500 mm/sec. The average diameter of the drops of metal did not exceed 5 mm. Consequently, the drop was transformed into a film in a time of 1/100 sec. If we take into account the fact that the heat is conducted through the upper and lower surface of the film, then even under the most unfavorable measuring conditions the cooling rate is of the order of 10^5 deg/sec. The temperature of the metal drop is 1700°C. If during the transformation of the drop into a metal film the temperature of the surface of the film drops 1000° then $1000 \times 100 = 10^5$ deg/sec. This figure is confirmed by measurements made in [167], where under much more favorable conditions it was found that the cooling rate was $3 \cdot 10^4$ deg/sec. The data on the rate of motion of the plates made it possible to determine also the average rate of crystallization of the alloy. The crystallization rate was calculated by the equation:

$$v = \frac{x v_{\text{u.p.}}}{2(D-x)} = \frac{0.1 \cdot 500}{2(5-0.1)} \approx 5 \text{ mm/sec},$$

where x is the thickness of the film; D the diameter of the drop; and $v_{\text{u.p.}}$ the rate of motion of the upper plate during the time the alloy hardens. The coefficient 2 is introduced because crystallization begins simultaneously on both sides.

The crystallization rate of cast iron was found to be ≈30 mm/sec.

The experiments on the preparation of films of rapidly cooled metals were duplicated by a simpler method in which a "catapult" was used (Fig. 38c). This apparatus consists of a steel lever 1 which is activated by a spring 2. A piece of the alloy 3 is placed in a cup at the end of the lever. The lever is placed in a horizontal position and the metal is melted in the furnace 4. Then the lever is freed and the metal is thrown onto the stationary copper plate 5. This method produces thin films of rapidly solidified metal. Co–C, Ni–C, and Fe–C alloys of the desired composition were smelted in a vacuum furnace [246]. The other alloys were smelted in graphite crucibles from quite pure initial materials, using the necessary fluxes.

The structure was examined by grinding off successive layers. The phase composition was determined by x-ray analysis. In some cases these data were supplemented by investigation with an electron microscope and volumetric analysis by the Saltykov method [178].

Let us first examine the results of the metallographic investigation of films of Fe–C alloys containing 2.1-5.5% C.

According to chemical analysis, the amounts of other elements present in the film did not exceed 0.1% P, 0.04% S, 0.4% Mn, and 0.3% Si.

In films of pre-eutectic composition there was usually separate crystallization. The structure of the alloy consisted of relatively large dendrites of austenite with thin cementite layers between austenite crystals (Fig. 39c).

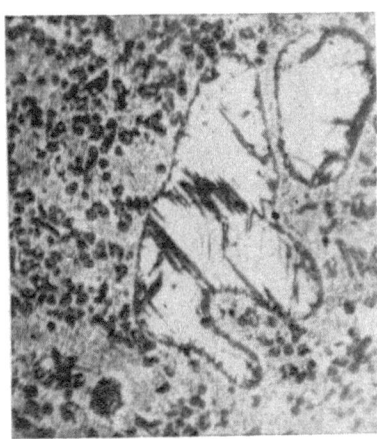

Fig. 41. Martensite in primary austenite. ×1000.

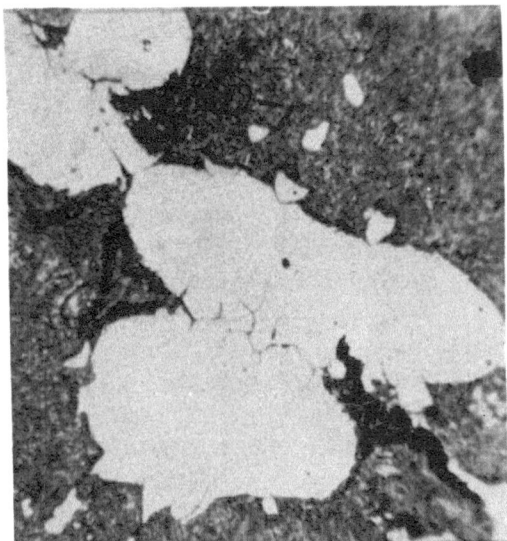

Fig. 42. Austenite at the surface of cast iron ("white" component). ×1000.

Spherulites of eutectoid colonies are often found on the surface of the film. As the composition approaches the eutectic composition, the degree of separate crystallization decreases and ledeburite colonies occur together with dendrites of austenite. The area occupied by austenite is larger than one would expect from the Fe–C phase diagram for films of this composition. In some cases the electron microscope revealed the fine structure of layers (resembling ledeburite) separating austenite crystals in alloys containing relatively low amounts of C (Fig. 40). The smaller the layers, the smaller the number of austenite inclusions in the ledeburite. This results from the fact that when the number of branched primary dendrites of austenite is large the layers of the remaining liquid become so small that the atoms of austenite in the eutectic liquid join the primary austenite crystals faster than the ledeburite can form. As the composition of cast iron approaches the eutectic composition, where the number of crystals of primary austenite decreases, ledeburite can crystallize, although in this case the primary austenite crystals are partially covered with layers of the austenite from the ledeburite.

Austenite grains contain a certain amount of martensite. It was found that the amount of martensite decreases with increasing grain size, apparently due to the degree of supercooling. The higher the degree of supercooling, the smaller the austenite grains. On the other hand, when the degree of supercooling is high the concentration of carbon in the austenite is lower,* and consequently there is more martensite (Fig. 41).

Austenite can crystallize in relatively large flat areas – the so-called "white" component – on the surface of films of pre-eutectic cast iron (Fig. 42) [231]. The number of such areas increases with decreasing concentrations of carbon in cast iron. Very small particles of a heterogeneous mixture are formed in immediate contact with crystallites of the "white" component. Quantitative analysis showed that the composition of this mixture is close to the eutectic composition.

Sometimes, thin layers of cementite occur between the austenite crystals and the heterogeneous mixture. Thus, the white austenite region has a composition different from the composition of the areas of the alloy surrounding it, which indicates that the components in the liquid phase are separated by diffusion during crystallization. The difference in the composition between the initial liquid and the heterogeneous mixture surrounding the crystals of the "white" component is particularly noticeable in alloys containing small amounts of carbon (2.2-2.4%), and it is in such alloys that one would expect crystallization without diffusion.

The results of microstructural analysis lead us to assume the following mechanism of the formation of the "white" component. In places where the contact between the liquid melt and the copper plate is very good, large austenite areas crystallize simultaneously. The liquid in contact with these areas becomes enriched in carbon. Since the heat is rapidly conducted away, the heterogeneous mixture and the cementite can crystallize immediately after the formation of the thin film of austenite.

---

* When the alloy is supercooled at a temperature below the eutectic line.

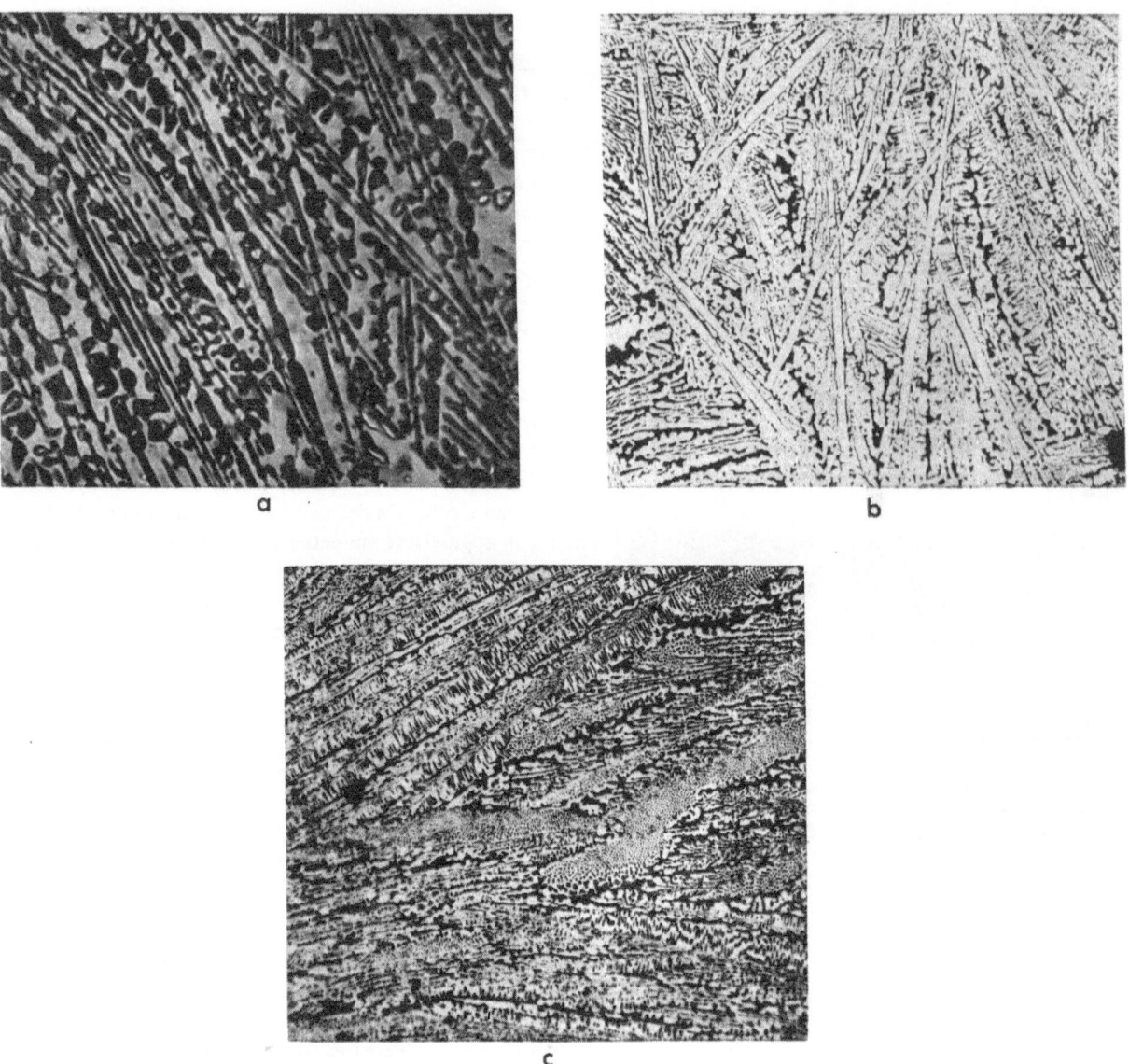

Fig. 43. Eutectic cast iron. Different cooling rates. a) At the surface; b) inside the ingot, ×500; c) in film. ×1500.

It is much more difficult to obtain separate crystallization of cementite and austenite in thin films of posteutectic alloys than in pre-eutectic alloys. Usually, one finds not only large plates of primary cementite but also ledeburite, apparently due to the fact that the cementite is not as branches as austenite, and the result is that the number of diffusion channels in the liquids increases and conditions for crystallization of ledeburite are improved.

Crystallization of cast iron of a eutectic composition is very interesting. Low-silicon cast iron containing 4.1-4.3% C has a eutectic structure when the cooling rate is very high [179]. An attempt to explain such a structure by "abnormalities" was made in [180].

A eutectic structure was found in [180] in an investigation of cast iron plates of eutectic composition obtained by pouring in thickwalled copper chill molds. In plates 5 mm thick only the middle of the plates had a eutectic structure (Fig. 43a). Plates of primary cementite near the surface are clearly visible (Fig. 43b).

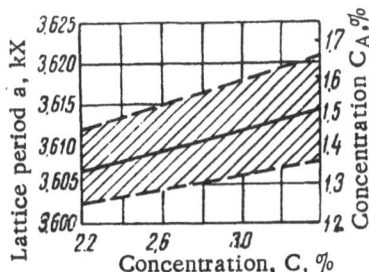

Fig. 44. Variation of the lattice constant of austenite and of its composition with increasing amounts of carbon in rapidly cooled molten cast iron. The width of the dashed part corresponds approximately to the difference in the concentrations of carbon in the outer and inner layers.

However, such an "abnormality" does not occur when the cooling rate is extremely high. In films of eutectic composition 0.05-0.2 mm thick solidified between two copper plates the structure is always eutectic with typical eutectic colonies (Fig. 43c). Thus, cementite plays an initiating role during the formation of the structure of eutectic cast iron only at intermediate cooling rates. At the very high cooling rate of about 10^5 deg/sec the cementite loses its initiating role in the formation of the eutectic structure in cast iron.

At intermediate cooling rates not only alloys of eutectic composition but also alloys of pre-eutectic composition have a tendency to form a eutectic structure [180]. At high cooling rates the situation is quite different.

A typical eutectic structure with characteristic ledeburite colonies without any traces of primary phases is formed only when the composition of the alloy is strictly eutectic.

X-ray analysis of rapidly cooled Fe–C films showed that their phase composition is the same as that of ordinary alloys quenched from high temperatures.

The cooling rates used stabilize the compositions of the phase formed during the crystallization of the alloy. Therefore, the measurements of the lattice constant of austenite should give information on the concentration of carbon in the primary phase of an alloy crystallized from a melt at a given temperature.

The lattice constant of austenite was measured by the method of inverted photographs with iron radiation. The (222) and (311) lines were used for the calculations. Pure copper was used as a standard. The precision of the measurements was 0.001 kX. Films of a pre-eutectic alloy containing 2.2, 2.5, 3.0, and 3.5% C were studied. Twenty-five films of each composition were used.

This investigation showed that the maximum amount of carbon in austenite, as determined by the value of the lattice constant, does not exceed 1.65%.*

The minimum concentration of carbon in austenite was 1.25% because the crystallization of austenite begins in the subcritical temperature range (below the eutectic line). The lower the crystallization temperature, the lower the limit solubility of carbon in austenite.

This assumption was confirmed by a study of the variation of the lattice constant of austenite with the distance from the surface of the film; austenite was examined after different amounts of the surface were ground off. It was found that the greater the distance from the surface the greater the lattice constant, which indicates that the concentration of carbon in austenite increases with the distance from the surface of the film (Fig. 44). The difference in the carbon concentration in the depth of the film and at the surface of the film is as high as 0.15% (absolute). This value is almost independent of the composition of the alloy.

Since the surface layers of the film are much more supercooled than the inner layers, one can conclude that the primary solid solution crystallizing from a liquid does not become supersaturated above the maximum solubility with increasing supercooling, but, on the contrary, its concentration decreases with increasing supercooling. Let us also note that the maximum solubility in Fe–C alloys usually means the concentration of austenite adjacent to cementite at the eutectic temperature (i.e., equilibrium with the metastable phase as distinguished from the stable equilibrium with graphite). The absence of highly supersaturated austenite in rapidly cooled Fe–C alloys was also noted in [182].†

---

* The Roberts equation [181] was used, since it is the most general of the equations in the literature.
†In this investigation the authors attempted to determine the limit solubility of carbon in austenite by the method of quenching from the liquid state. As we have said, this method cannot be used for this purpose because the increase of the degree of supercooling in the subcritical temperature range leads to a decrease in the concentration of the primary solid solution.

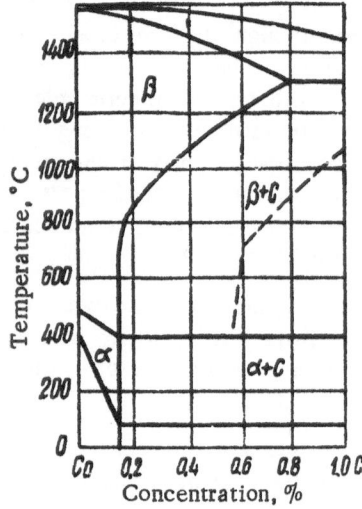

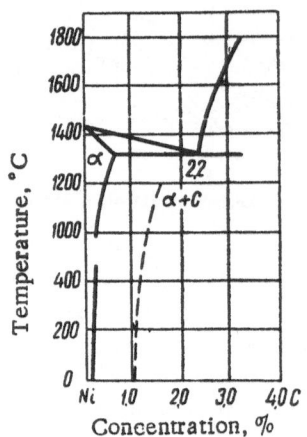

Fig. 45. Phase diagram of the Co—C alloy. Solubility of the α-solid solution in the case of metastable equilibrium.

Fig. 46. Phase diagram of Ni—C. Solubility limit in the case of metastable equilibrium.

Let us now examine the results of the investigation of the structure of films of Co—C and Ni—C alloys. Metallic cobalt exists in two modifications. The low-temperature modification (α-Co) has a hexagonal close-packed lattice. The high-temperature modification (β-Co) has a face-centered close-packed lattice [183]. The α →β transformation occurs at 420°C.

The phase diagrams of Co—C and Ni—C alloys are shown in Figs. 45 and 46. One can see that the maximum solubility of carbon in the primary solid solution at the eutectic temperature is 0.8-1.0% (by weight) for Co [184, 185, 189] and is 0.55-0.65% for the primary solid solution based on Ni [184-186]. Even at relatively high rates of cooling from the liquid state, the author could not prevent the formation of graphite in alloys in which the concentration of carbon exceeded the maximum solubility at the eutectic temperature (C_m) [183, 184]. In [187] the authors observed the presence of white crystals of a new phase which was assumed to be the cobalt carbide Co_3C in the Co—C alloys and the Ni_3C carbide in the Ni—C alloys. The same carbides were found when pure cobalt and nickel were carburized [184].

Metallographic investigations of Co—C and Ni—C films showed that separate crystallization occurs when the cooling rates are very high. The cobalt and nickel carbides are metastable phases. It is quite easy to obtain the carbide eutectic in Co—C alloys, but the eutectic occurs much less often in Ni—C alloys. Among the Fe_3C, Co_3C, and Ni_3C carbides, nickel carbide is the least stable. In alloys containing a large amount of carbon (about 3% and higher) globular graphite occurs along with the carbide, and the amount of globular graphite increases with the distance from the surface of the film. The thinner the film the less graphite it has; there is almost no graphite in thin films.

Metallographic analysis showed that the eutectic point at which the heterogeneous mixture of the solid solution and the carbide is formed is shifted toward higher concentrations of carbon. For example, the structure of the Co—C alloy containing 4.4% C is pre-eutectic. The structure of the Ni—C alloy containing 3.5% C is also pre-eutectic.

X-ray analysis of these alloys revealed the following. No polymorphic transformation occurs in films of Co—C alloys. The x-ray photographs show only the lines of the high-temperature modification (β-Co). The maximum carbon concentration in the β-solid solution of the Co—C alloy is 1.65%. In Ni—C alloys the maximum solubility reaches 1.85%. The variation of the lattice constant of the solid solution of these alloys is shown in Fig. 47.

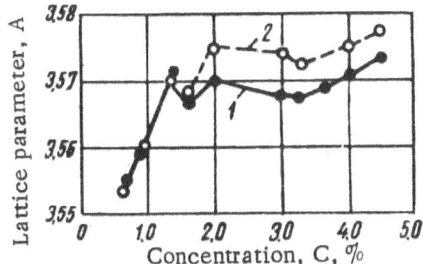

Fig. 47. Variation of the lattice parameter of the α-solid solution in the Co–C alloy with the concentration of carbon. 1) In the outer layer of film; 2) in the inner layer of film.

The data on the concentration of carbon in the primary solid solution do not make it possible to represent the equilibrium conditions in Co–C and Ni–C systems by double phase diagrams. The solubility limit in a metastable system is shown by the dashed lines in Figs. 45 and 46.

If the concentration of carbon in an alloy does not exceed the maximum solubility for a metastable system (C_m) then as a rule the alloy consists of a single-phase supersaturated solution. X-ray photographs of the Co–C alloy showed that cobalt carbide lines appear when the carbon concentration increases. A further increase in the concentration leads to the appearance of the graphite line. X-ray photographs often show two systems of lines relative to the lattice of the primary solid solution. One pair of lines corresponds to the section of the solid solution adjacent to the graphite, where the concentration is low, and the other refers to the section of the solid solution in equilibrium with the carbide, where the carbon concentration is high (Fig. 48).

In this case, also, the lattice constant of the primary solid solution varies from the surface to the middle of the film. As a rule, the solid solution contains less carbon at the surface than in the middle of the film. This result is without doubt the consequence of the precipitation of graphite on the surface observed in films with high carbon concentrations. The same situation occurs in layers with no graphite on the surface (with a relatively low concentration of carbon).

Metallographic investigations of the other alloys (Al–Mn, Al–Cr, Al–Si, Al–Zn, Pb–Sn, and Bi–Sn) showed about the same tendency of pre-eutectic alloys toward separate crystallization and of eutectic alloys to form eutectic colonies in the case of strictly eutectic compositions.

X-ray analysis showed that the primary solid solution in the Al–Mn alloy becomes supersaturated above the maximum solubility [168]. The minimum lattice constant of the solid solution of Mn in Al is 4.011 kX; according to the Vegard law this lattice constant corresponds to a concentration of 9% Mn (by weight); the maximum solubility under equilibrium conditions is 1.4%. Under these conditions a metastable phase of another composition is formed instead of the stable Al_6Mn phase. The low absorption capacity of Al prevented the authors from determining the variation of the concentration of Mn in the solid solution at different distances from the surface.

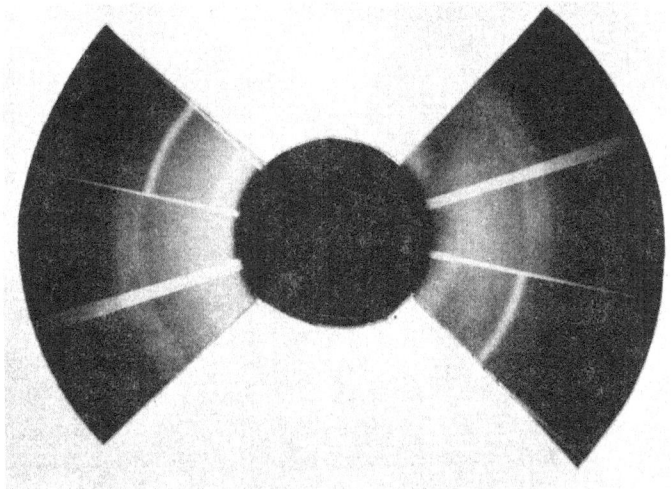

Fig. 48. X-ray photographs of Ni–C containing 2% C. The (311) line is double. The (311) line of the Ni standard is next to it.

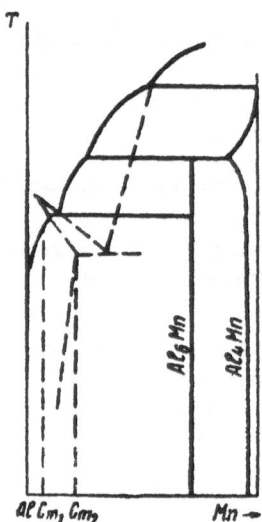

Fig. 49. Phase diagram of Al—Mn.

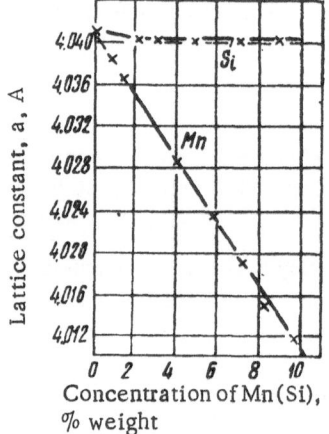

Fig. 50. Variation of the lattice period of the α-solid solution in Al—Mn and Al—Si alloys as a function of the concentration of the second component.

This alloy can be represented by a double phase diagram (Fig. 49). The variation of the lattice constant is shown in Fig. 50.

In Al—Cr alloys the solubility of Cr in Al is also higher than the maximum solubility when the alloy crystallizes under equilibrium conditions. These data are in complete agreement with the results obtained in [168].

X-ray analysis of Al—Si, Al—Zn, Pb—Sn, and Bi—Sn alloys showed that primary solid solutions with concentrations exceeding maximum solubility at the melting temperature of the eutectic do not form at high cooling rates. In Al—Si alloys the maximum solubility of Si in Al is 1.65% at 578°C. According to [168], the minimum lattice constant of Al saturated with Si atoms is 4.0409 A. (The lattice constant was determined in alloys quenched from high temperature in the solid state.) When the cooling rate from the liquid state is high (using either of the two methods previously described) the lattice constant of the primary solution of Si in Al is never lower than 4.0410 A regardless of the concentration of the alloys (up to 15% Si). The Debye diagrams of these alloys and alloys of the same composition from the solid state are the same. The average of fifty determinations of the lattice constant of the primary solid solution of Si and Al corresponds to the solubility of ~1.4% (Fig. 50).

In Pb—Sn and Bi—Sn alloys the lattice constant of primary solid solutions (α and β) was determined on films with compositions varying from the limit solubility of the α-solid solution to the limit solubility of the β-solid solution at intervals of 10%. The lattice constant of alloys of the same composition quenched in the solid state from a temperature close to the eutectic temperature was also measured. In no case was the lattice constant of films greater than the lattice constant of the primary solid solutions quenched from high temperatures in the solid state. The authors did not succeed in obtaining supersaturation above the maximum solubility at the eutectic temperature either in α- or β-primary solid solutions.

The average of fifty determinations of the lattice constant of Pb—Sn films was 4.9395 A for the solid solution of Sn in Pb; in the solid solution of Pb in Sn a = 5.377 A and c = 2.882 A.

In Bi—Sn alloys the maximum value of the lattice constant of the solid solution of Bi in Sn was a = 5.827 A and c = 3.186 A. Debye diagrams of pre-eutectic and posteutectic films of these alloys always contained lines of the solid solutions of both components. In alloys of eutectic composition the dispersity of the structural components is so high that it is difficult to determine the lattice constants of the solid solutions composing the eutectic. This statement refers to Fe—C, Co—C, Ni—C, Al—Mn, and other alloys.

In Al—Zn alloys primary solid solutions with maximum concentrations of Zn as high as 65% by weight were stabilized by quenching from the liquid state. (The phase diagram of the Al—Zn system is shown in Fig. 51.) The values of the lattice constants agree with those obtained in [275]. The variation of the lattice constant found in [275] is shown in Fig. 52.

Further increase in the concentration of Zn in the initial melt results in the fact that a solid solution of Al in Zn not exceeding the maximum solubility is always present in quenched films. The precipitation of Zn from alloys containing more than 65% Zn is not stopped even by quenching of films from the temperature of the homogeneous solid solution (t = 400°C). This result indicates that the rate of formation and growth of crystals during the decomposition of highly supersaturated solid solutions is extremely high. Therefore, the decomposition rate of a highly supersaturated liquid solution must be even higher.

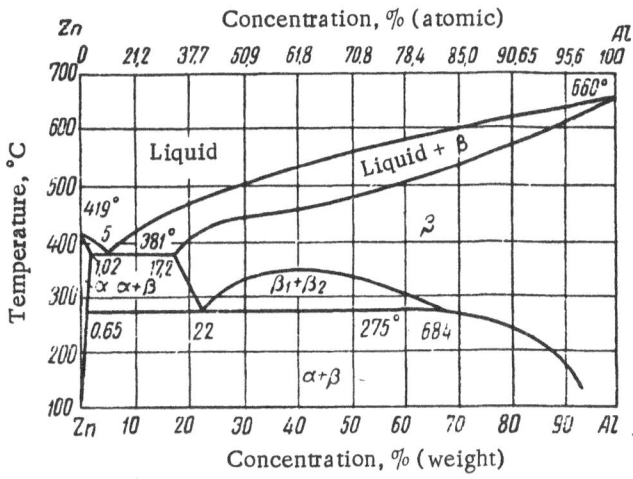

Fig. 51. Phase diagram of Al—Zn.

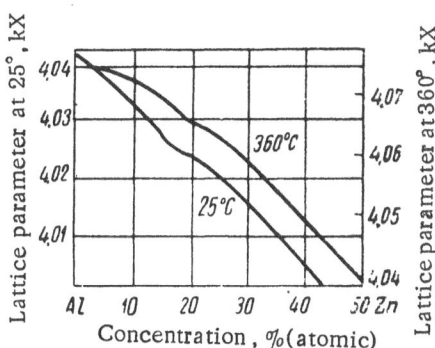

Fig. 52. Variation of the lattice constant of the aluminum base α-solid solution as a function of the concentration of Zn[275].

4. Singularities of the Crystallization Process of Alloys at High Degrees of Supercooling

Unfortunately, the methods used for very rapid cooling make it impossible to record the temperature during the transformation of the alloy from the liquid to the solid state because the volume of solidifying material is so small. It should be noted that in the final analysis the most important matter is to determine the difference in temperatures of the crystallization front and the liquid metal. At the present time such measurements are almost impossible in the more complex cases of crystallization. As the result, the real degree of supercooling remains unknown. It is known only that the methods described here provide the highest degree of supercooling during the growth of crystallites as compared to other methods. Crystallization centers can occur in other methods of high degrees of supercooling because of the homogeneous formation of nuclei. However, during the growth of nuclei the heat of crystallization increases the temperature at the crystallization front considerably. Since the chemical composition of the phases is determined by the composition of the crystallites (and not the nuclei), the main factor is apparently the degree of supercooling reached during the growth of the crystal.

Judging by the structure of alloys formed at high degrees of supercooling, all the systems investigated can be divided into two groups:

1. Systems with unlimited solubility in the solid and liquid states. This group also includes the primary solid solution in which the concentration does not exceed the maximum solubility at the melting temperature of the eutectic (peritectic). The data on the effect of high cooling rates on the formation of the structure in these alloys indicate only that there is no intercrystalline liquation in the final structure, although rather precise methods for local analysis have been used for such investigations.

2. Systems with unlimited solubility in the liquid state and limited solubility in the solid state not forming intermediate phases. Al—Si, Al—Zn, Pb—Sn, Bi—Sn, etc., belong to this group. When the degree of supercooling is high these systems do not form highly supersaturated solutions with a solubility exceeding the maximum solubility at the eutectic temperature. In most cases the solubility of primary solid solutions remains below C_m with increasing rates of cooling from the liquid state.

3. The Cu—Ag eutectic system does not form intermediate phases and is the only system known so far with a diffusionless transformation. According to [170], the cooling rate necessary for this transformation is

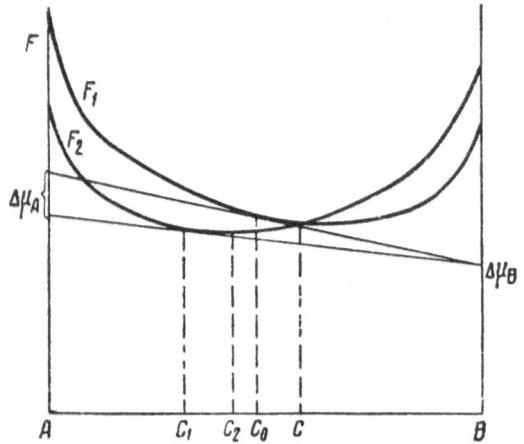

Fig. 53. Diagram of the variation of the free energy with the composition (diffusionless transformation in a cigar-type system).

close to 10^5 deg/sec. It should be emphasized that the Cu–Ag system satisfies all the requirements (listed in Section 3, Chapter 1) for forming a continuous series of solid solutions.

4. Systems with unlimited solubility in the liquid state and limited solubility in the solid state which form intermediate phases. This group includes Al–Mn, Al–Cr, Al–Ti, Al–V, Al–Fe, Pb–Na, Pb–Te, Pb–Cu, Al–Cu, Pb–Ag, Fe–Zn, Co–Zn, Ni–Zn, Fe–C, Co–C, Ni–C, Hg–Na, and many others. During crystallization of these alloys at high degrees of supercooling, highly supersaturated solid solutions occur in which the concentration is much higher than the maximum equilibrium concentration at high temperature. However, there is also a maximum concentration C_m in these systems above which it is impossible to supersaturate the solid solution by supercooling alone. If the concentration of the alloy exceeds C_{m_1} then a two-phase metastable system is formed and this metastable system contains an intermediate metastable phase. The composition and melting point of the metastable phase are different from those of the stable phase. Thus, in the Al–Mn system the melting point of the metastable phase and the concentration of manganese are much higher than in the stable phase. In the Co—C system the melting point of the Co_3C metastable phase is lower than that of the stable phase and the amount of carbon in the metastable phase is much lower than the amount of graphite in the stable phase.

Let us analyze these experimental data by applying the thermodynamics and kinetics of phase transformations. Let us first examine crystallization from highly supercooled cigar-type systems. Let us assume that the degree of supercooling is such that the nuclei begin to occur at a temperature only a few degrees above the solidus temperature. The free energy curves relative to this case are shown in Fig. 53. The $F_1(C)$ curve corresponds to the liquid phase, the $F_2(C)$ curve to the solid phase; the composition of the crystallizing liquid solution is C_0.

At the present time, there are no data which could be used to determine precisely whether the composition of the crystallization centers of the precipitating solid solution is different from the composition at which the condition $\Delta\mu_A = \Delta\mu_B$ is satisfied. Apparently, new phases can precipitate, capturing atoms of the component in which the gradient of chemical potential is smaller than that of the solvent. When the concentration of such a component is low its atoms can be attracted by the atoms of the surrounding solvent. This statement is illustrated by the diagram in Fig. 53. Let us assume that the composition of the new phase precipitating from the solution is different from the composition responsible for the equality of the chemical potentials of phases, so that $\Delta\mu_A > \Delta\mu_B \geqq 0$. This means that the forces which attract the A atoms to the growing center of the new phase are larger than the forces acting on the atoms of the B component in the same direction. When the conditions are such that the mobility of the atoms is low, the nucleus can grow until it reaches the composition C_1. Beyond C_1 the forces acting on the atoms of the B component are in the opposite direction. Since the forces of atomic interaction are effective only at very short distances, it is difficult to assume that the precipitating phase can capture and attract the atoms of the B component. According to the diffusionless crystallization hypothesis, the atoms pass from one phase into another by just this capture mechanism. The only condition for the existence of diffusionless passage is that the free energy of the mother phase F_1 be greater than the free energy of the precipitating phase F_2.

Thus, in Fig. 53 point C corresponds to the composition of phases at which the free energy of the first phase is equal to the free energy of the second phase; therefore, for any composition of the first phase to the left of point C the precipitating second phase has the same composition as the liquid (for example, C_2), since $F_1 > F_2$.

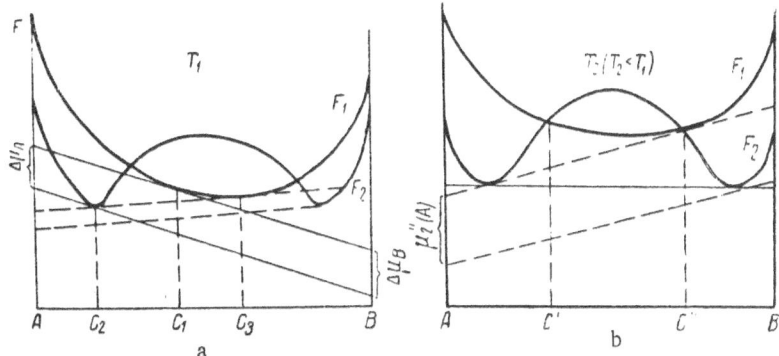

Fig. 54. Diagram of the variation of free energy with the composition
(diffusionless transformation in a cigar-type system).

It should be noted, however, that according to the phase diagrams of systems with unlimited solubility in the liquid and solid phases, the melting point of all the crystals of the solid solution with a composition corresponding to the points between C_0 and C_1 are lower than the melting points corresponding to this degree of supercooling, and as the result crystals cannot form from liquids. Therefore, in this case diffusionless crystallization from the liquid state is possible only when the melt is supercooled below the solidus temperature.

Investigators trying to show the possibility of diffusionless transformation of the liquid into the solid state when the liquid is supercooled below the temperature at which the thermodynamic potentials are equal should have observed the solidification of the alloy at a temperature above the solidus temperature. However, this phenomenon does not occur in systems which do not form intermediate phases. The unique experiment (by D. S. Kamenetskaya) in which this phenomenon did occur concerned the Hg–Na system, which does form intermediate phases. We shall discuss this experiment later.

For the present, there is no reason to relate diffusionless crystallization in systems with unlimited solubility to curves of equal free energies. It is simpler and qualitatively more correct to relate diffusionless crystallization to the gradient of chemical potentials of the atoms of the components. If the alloy is supercooled below the solidus temperature and $\Delta\mu_i$ are unequal but positive, then diffusionless crystallization occurs with capture of the atoms of the B component in the second phase.

Let us now examine the crystallization process in systems with limited solubility in the solid state, which includes alloys with eutectic and peritectic transformations. The $F_1(C)$ and $F_2(C)$ curves are shown in Fig. 54a. These curves describe the dependence of the free energies of the liquid and solid phases on the concentration for a given degree of supercooling. Let us assume that the composition of the supercooled liquid corresponds to point C_1. Then the composition of the precipitating primary solid solution will be C_2, since the slope of the tangent at this point is the same as the slope of the tangent at point C_1. Here $\Delta\mu_A = \Delta\mu_B$. During precipitation, the liquid is impoverished in the A component and is enriched in the B component. At the same time, the α-primary solid solution also becomes saturated with the B component. But the chemical potentials of the phases remain equal and the gradients of the chemical potentials of the components decrease continuously but remain equal: $\Delta\mu_A = \Delta\mu_B$. Finally, at some stage – let us assume when the liquid reaches concentration C_3 – the gradient of chemical potentials of the atoms of the components in the primary β-solid solution becomes sufficient for the formation of the nuclei of the phase. Then, the growth of the nuclei of β-phase is more advantageous than that of the nuclei of the α-phase and the liquid is impoverished in the B component up to the point where the gradients of the chemical potentials of the atoms of the components of the α- and β-phases reach equilibrium. From this moment onward there is a two-phase decomposition of the liquid with precipitation of the eutectic mixture. The compositions of the phases in this mixture correspond to the equilibrium concentrations of primary solid solutions at this temperature.

With increasing degrees of supercooling the distance between the minima on the $F_1(C)$ and $F_2(C)$ curves increases, since the limit solubilities of the primary solid solutions decrease with decreasing temperature.

Figure 54b shows the same curves as those in Fig. 54a, but at a higher degree of supercooling. It can be seen that the radius of curvature increases and the distance between the minima on the two curves increases. In this case (for a given pre-eutectic composition of the liquid C_1) the composition of the precipitating α-phase is below C_2 and the nuclei of the β-phase will begin to precipitate at a lower degree of supersaturation of the liquid; consequently, the degree of supersaturation of the α-phase will be lower. After the nuclei of the β-phase are formed they will grow simultaneously with the crystallites of the α-phase and an equilibrium will be established corresponding to the minimum gradient of chemical potentials of the atoms of the A and B components in the solid and liquid phases. This condition is satisfied when the concentrations become equal to those determined by the common point of tangency of the $\mu_2(A)$ and $\mu_2(B)$ lines. Thus, an increase of the degree of supercooling cannot increase the solubility of the primary solid solutions above C_m, since the second phase is precipitated before the α-phase reaches the limit solubility at this temperature.

Very often, to explain such phenomena as the formation of abnormal structures (quasi-eutectics) and the crystallization process with seed crystals above the metastability limit of the alloy (i.e., at a relatively low degree of supercooling) one assumes that the primary solid solution can be supersaturated above C_m. However, there have not been any direct experiments or theoretical analyses which show the existence of a highly supersaturated phase in such a system. Therefore, at present we may assume as proven that the increase of the cooling rate in systems which decompose during crystallization from the liquid state decreases the concentration of the primary solid solution and does not increase it (above C_m).

Let us now return to the hypothesis of diffusionless crystallization by capture of atoms possessing a lower gradient of chemical potentials (i.e., atoms moving from phase I to phase II under the effect of forces which are weaker than the forces acting on the atoms of the main component). We must examine this hypothesis because transformations of the martensitic type cannot be excluded from a general study of crystallization at a high degree of supercooling.

Since under conditions in which the mobility of atoms is low the establishment of mechanical equilibrium lags behind diffusionless crystallization, which occurs very rapidly, we may assume that such a process is possible at much higher degrees of supercooling than those used in our investigation. The curves of free energies of the liquid and solid phases referring to this case are shown in Fig. 37b. When liquid with a composition C_3 crystallizes, the atoms of the A and B components pass from the liquid to the solid phase of the same composition at a positive, although unequal, difference of chemical potentials of the components because the distance between curves I and II is very large. Consequently, diffusionless crystallization of the martensitic type can occur.

However, experimental data on crystallization in a large number of eutectic systems do not confirm this assumption.

Why then can a system in the solid state be transformed without diffusion into a highly supersaturated solid solution, while in the liquid state it cannot be?

The simplest answer to this question is that the mobility of atoms in the liquid is so high that the mechanical equilibrium between atoms always occurs before groups of atoms pass from the liquid into the solid state. However, diffusionless crystallization apparently does occur in highly supercooled systems with phase diagrams of the cigar type. On the other hand, there are cases of the supercooling of liquids up to the point of transformation into the amorphous state (without phase transformation).

Therefore, the following hypothesis is worth considering: If a liquid solution is simultaneously supersaturated with two components then the free energy of the solution is always lower than that of the solid solution of the same composition. From the geometric viewpoint, this means that the free energy of the liquid phase is lower than the free energy of the solid phase regardless of the degree of supercooling in the range of concentrations from C' to C" (Fig. 54b).

The physical meaning of this assertion is that supersaturation of solid solutions is accompanied by an increase of internal stresses resulting from the deformation of the crystal lattice and this increase affects the value of the mixing energy. The analysis of the equations given in Chapter 3, taking into account the dependence of the mixing energy on the temperature, confirms the validity of this assertion.

Let us examine the crystallization conditions in different systems at high degrees of supercooling.

In a system with unlimited solubility in the liquid and solid states (cigar type) diffusionless transformation can occur when the degree of supercooling is sufficiently high and the transformation temperature is below the solidus temperature. Then both the condition $\Delta\mu_A > \Delta\mu_B \geq 0$ and the condition $\Delta\mu_A = \Delta\mu_B > 0$ are satisfied. From the geometric viewpoint, this means that all the points of the $F_1(C)$ curve are above the $F_2(C)$ curve (Fig. 53). It is easy to show that this is actually possible for a system of this type.

In a system of the cigar type the mixing energies of the liquid and solid phases are $V' = V'' < 2kT$.

From the equations of the type of (2.6) it follows that when $C_1 = C_2$ we have

$$F_1(C) - F_2(C) = NC(1-C)(V'-V'') + (1-C)Q_A + CQ_B - T(1-C)\frac{Q_A}{T_A} - T\frac{Q_B}{T_B}, \qquad (6.1)$$

where Q_A and Q_B are the heats of fusion of the components and T_A and T_B are the melting points if:

$$F_1(C) - F_2(C) = (1-C)Q_A + CQ_A - T\left[(1-C)\frac{Q_A}{T_A} + C\frac{Q_B}{T_B}\right].$$

Consequently, the distance between the $F_1(C)$ and $F_2(C)$ curves increases for any concentration C with an increasing degree of supercooling (decrease of T).

Let us now investigate a system with unlimited solubility in the liquid state and limited solubility in the solid state which does not form intermediate phases. In this case one would expect diffusionless crystallization if one could reach a degree of supercooling at which all the points of the $F_1(C)$ curve are much higher than those of $F_2(C)$ (Fig. 54). Then, for many concentrations not only the condition $\Delta\mu_A > \Delta\mu_B \geq 0$ but also the condition $\Delta\mu_A = \Delta\mu_B > 0$ would be satisfied.

For this system the mixing energy usually satisfies the condition $V' < 2kT$ and $V'' > 2kT$, and in any case $V' < V''$.

Equation (6.1) shows that if $\Delta V'$ is independent of temperature then all the points of the $F_1(C)$ curve are above the $F_2(C)$ curve when the temperature decreases. However, an investigation showed that the mixing energy of the solid solution increases with decreasing temperature much faster than the mixing energy of the liquid phase [48, 49]. According to these investigations, the mixing energy of the solid solution $V'' = V_0' + \varepsilon$, where ε is the additional energy proportional to the modulus of hydrostatic compression. The dependence of the modulus of hydrostatic compression on the temperature [190] can be represented in first approximation as $\varepsilon = \varepsilon_0 - AT$, where A is some coefficient, and then $V'' = V_0'' - AT$, and consequently:

$$F_1(C) - F_2(C) = NC(1-C)\left[(V'-V_0'') + AT\right] + (1-C)Q_A + CQ_B - T\left[(1-C)\frac{Q_A}{T_A} + \frac{Q_B}{T_B}\right],$$

i.e., ΔF tends to decrease instead of increasing after having reached a certain value. This conclusion was confirmed experimentally. In such systems the increase of the cooling rate leads to the decrease of the concentration of the primary solid solution below C_m. Apparently supercooled liquids with a concentration above C_m have a lower free energy than the solid phase at the same concentration regardless of the temperature.

Let us also determine why the martensitic transformation is possible. This transformation occurs in the case of supercooling in the solid state. In this case the mixing energies of the mother solid solution V' and the forming martensite phase V'' increase with decreasing temperature, and consequently ΔV is almost independent of temperature in some alloys in which the martensitic transformation is possible. In this case both the $F_1(C)$ and $F_2(C)$ curves can be situated so that $\Delta\mu_A = \Delta\mu_B > 0$ or $\Delta\mu_A > \Delta\mu_B \geq 0$ (Fig. 55). Obviously, this situation requires some critical degree of supercooling which, by the way, depends on the composition of the alloy. Also,

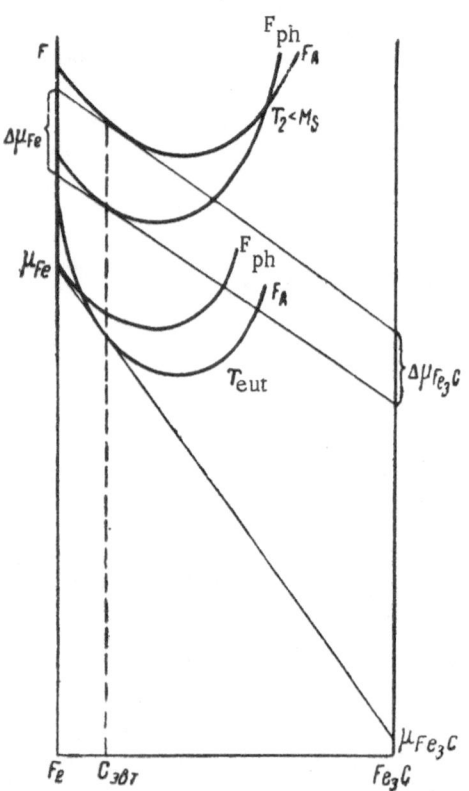

Fig. 55. Diagram of the variation of free energy with the concentration of carbon (martensitic transformation).

the precipitation of the martensite itself leads to volume changes as the result of which V' and V" change in such a way that further precipitation of martensite requires a decrease in the temperature.

Thus, diffusionless transformation of supersaturated solutions can occur in the following three cases:

1. In systems with unlimited solubility in the liquid and solid states supercooled below the solids temperature.

2. In the solid state, if the difference between the mixing energies of the two phases does not increase with increasing degrees of supercooling.

3. In a system with unlimited solubility in the liquid state and limited solubility in the solid state provided the supercooling is below the solidus temperature or that the difference between the mixing energies of the liquid and solid states does not increase with increasing degrees of supercooling. This latter case occurs very seldom and refers to systems similar to Cu–Ag, i.e., formed by components with similar physicochemical properties (small difference between atomic diameters, lattice of the same type, and same valence). In such systems the difference between the mixing energies is relatively small, and consequently the curve representing the variation of the free energy with the concentration has very weak minima. Calculations show that $\Delta V = 5693$ cal/mole − 3193 cal/mole = 2500 cal/mole. This is a very small value compared to $\Delta V = 1139$ for the Bi–Sb system, which forms solutions with unlimited solubility in the solid and liquid states. The free energy curves for the Ag–Cu system are shown in Fig. 15 (p. 22).

The large number of experimental data relative to the laws of formation of the structures of alloys at high cooling rates cannot be generalized by the theory of diffusionless crystallization in the form given in [159,160]. These data can be generalized at least qualitatively on the basis of thermodynamic concepts and the molecular-kinetic theory of phase transformations, taking into account the dependence of the mixing energies on the transformation temperature. It should be remembered that the transformation temperature is determined not only by the dependence of the modulus of hydrostatic pressure on temperature but also by the valences and the difference in the lattices of the components of the alloy, the activity, the presence of various defects, etc.

But what is the physical meaning of the lines of equal free energies (abc in Fig. 36)?

This curve indicates that at temperatures below T_1 the solid phase in which the concentration of the second component is lower than C_1 has a lower free energy than the liquid phase of the same composition. This, however, is true only for isolated phases (the liquid and the solid phases separately), at the boundary with a vacuum, for example.

At the moment these phases come in contact, the equilibrium is disrupted and the concentrations of the components are redistributed. It follows, then, that the physical meaning of the curve of equal free energies is that it indicates the limit of the temperature stability of a homogeneous solid solution with a given composition during superheating. In a unicomponent system this limit is the melting point. The solid solution can be superheated (without melting) above the solidus line up to the line of equal free energies; after reaching this line the solid solution melts almost instantaneously − without any diffusion. Obviously, when the homogeneous solid solution is kept for a long period of time below the points of equal free energies (above the solidus) it may begin to melt, but it will melt without diffusion. In this case the phases separate, with simultaneous redistribution of concentrations according to the equilibrium phase diagram.

Consequently, the line of equal free energies is the geometric locus of different points of diffusional melting of a homogeneous solid solution at high heating rates.

All these conclusions follow from the investigation of the formation of the structure of alloys with unlimited solubility in the solid and liquid states [303, 304].

5. Transformations in Systems with Intermediate Phases. Double Phase Diagrams

In alloys forming intermediate phases supersaturated primary solutions with a concentration above C_m can be formed even when only one primary solution exists in the solidified system. Thus, alloys of Ni–C, Co–C, Al–Mn, etc., with concentrations exceeding C_m (sometimes by a factor of 10) crystallize in the form of primary solid solutions of the same composition when the degree of supercooling is sufficiently high.

At first glance this phenomenon confirms the necessity of extending the solidus line to the subcritical region, and consequently contradicts the conclusions derived in the preceding sections. If, however, we analyze the process of formation and growth of nuclei in these systems it turns out that the formation of highly supersaturated solid solutions is the consequence of the same laws, which, however, are complicated by the presence of intermediate phases.

Examination of the phase diagrams of these systems shows that intermediate phases act as independent components and divide the diagram into a series of independent phase diagrams. It is clear that these components are not independent and therefore cannot be used to draw multicomponent phast diagrams similar to those of ternary or quaternary systems.

The reason for this is that the intermediate phases are formed by the same initial components. Therefore, according to the Gibbs phase rule, the intermediate phase cannot be in equilibrium with two initial phases (containing the atoms of primary components) in a two-component system.

In iron-carbon alloys (not containing any other elements) the austenite (ferrite), cementite, and graphite cannot be in equilibrium. Here, the cementite is in equilibrium either with the graphite or with the austenite (ferrite). Nevertheless, the cementite in equilibrium with each of the main phases acts as an independent component, and therefore the diagram appears to consist of independent phase diagrams. However, on the basis of experiments, diagrams of metastable equilibrium are drawn in dotted lines in many cases (Fe–C, Fe–P, etc.). The analysis of the results of the modern thermodynamic theory of the construction of phase diagrams shows that equilibrium in systems forming intermediate phases must be described by double phase diagrams. One of these diagrams, corresponding to a low cooling rate, describes the stable equilibrium, and the second, relative to a high cooling rate, indicates the equilibrium of metastable phases.

In fact, the dependence of the free energy of phases on the concentration given by the equation

$$F(C) = NC(1-C)V + (1-C)V_A + CV_B$$

$$+ kT \left\{ C \ln C + (1-C) \ln(1-C) - \frac{T}{N}[(1-C)S_A + CS_B] \right\}$$

is determined by the properties of the components forming the alloy: here, V_A' and V_B' are binding energies, and S_A and S_B are the entropies of the components of the respective phases. If we assume that the intermediate phase behaves as an independent component, then it forms, with other components, phases which have independent free energy curves.

When an ordinary liquid solution is supercooled, the atoms form a definite close order which (together with other factors) determines the dependence of the free energy on the concentration of the liquid. The probability of the formation of a new phase depends greatly on the type of the close order. In ordinary systems (not forming intermediate phases) the structure of close order in the liquid is not changed by increasing degrees of supercooling. If, however, the system does form an intermediate phase, then the close order promoting the precipitation of this phase occurs in a highly supercooled liquid in which the concentration is very different from that at which the intermediate phase precipitates under stable conditions. The increase of the degree of supercooling decreases the kinetic energy of the atoms of the liquid, and consequently increases the stability of

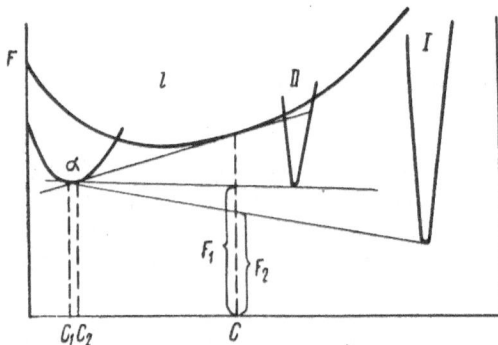

Fig. 56. The concentration of the metastable phase is lower that that of the stable phase.

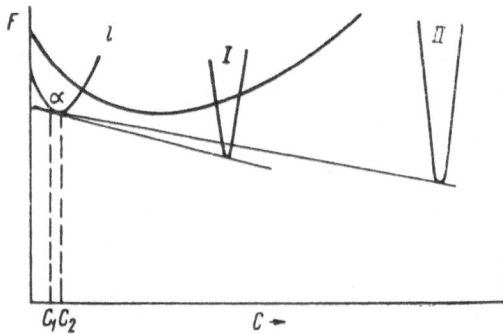

Fig. 57. The concentration of the metastable phase is higher than that of the stable phase.

different associations of atoms in which the order is close not only to the stable phase but also to the metastable phase. In the second case the dependence of the free energy of the liquid on the concentration is described by a function which includes other constants [192].

In those regions of the supersaturated liquid solution where the metastable close order is retained the precipitating solid phase has a much higher limit solubility than in the case of stable order. This results from the fact that the solubility of the solid phase depends not only on the properties of the phase itself but also on the properties of the surrounding mother phase. Therefore, the solidus and liquidus lines, whose position is determined by the equilibrium conditions of the liquid and solid phases in systems forming intermediate (metastable) phases, must be double. For a given system they occupy definite positions, which determine the equilibrium in stable and metastable systems. This assumption was confirmed by D. S. Kamenetskaya's experiments in which the Hg–Na alloy was completely transformed into the solid state above the solids line. Hg–Na alloys form intermediate phases. Two primary solid solutions with different lattice constants were found in Ni–C alloys. Each of these solid solutions is in equilibrium with one of the two phases and the solid solution with higher solubility is in equilibrium with the metastable phase.

Figure 56 shows hypothetical curves of the variation of free energy with the concentration in the case where the metastable phase II with a concentration lower than that of the stable phase I is formed in a binary system. In this case phase II is in equilibrium with the solid solution whose concentration C_2 is higher than the concentration of the stable phase I (C_1). The degree of metastability of the system containing a metastable phase is determined by the relative level of the free energy of the whole system. Thus, for an alloy with a concentration C the free energy of the metastable state is indicated by segment F_1, while for the stable state it is given by segment F_2.

Figure 57 is a diagram of the case where the metastable phase in the system has a higher concentration than the stable phase. The α-solid solution at the boundary with phase II is then supersaturated. Supersaturation can greatly exceed C_{max}, i.e., the maximum concentration in the case of the equilibrium between the α-phase, phase I, and the liquid at the eutectic temperature. However, in this case also there is a maximum possible solubility of the α-phase (C_{m_2}) which depends on the properties of the metastable phase adjacent to it. The concentration of the α-phase will be lower than C_{m_2} when the degree of supercooling increases.

In the case discussed here the metastable phase is the phase whose concentration is higher than that of the primary solid solution. But this statement does not mean that the metastable phase should have a higher free energy than the stable phase. If we represent the geometric locus of points of equilibrium between the stable and metastable phases and the primary solid solution in the form of an equilibrium diagram we obtain a double phase diagram similar to those used to describe the equilibrium in alloys of Fe–C, Fe–P, etc. Figure 57a and b, shows two types of double-phase diagrams which describe phase equilibrium in systems forming intermediate phases at high degrees of supercooling.

These phase diagrams can be constructed by using experimental data obtained as the result of determining the composition of phases of systems crystallized at high cooling rates. In these systems the highly supersaturated solid solution results not from diffusional crystallization but from the crystallization of metastable phases. The precipitation of metastable phases is explained by the creation of favorable kinetic conditions for the formation and growth of nuclei.

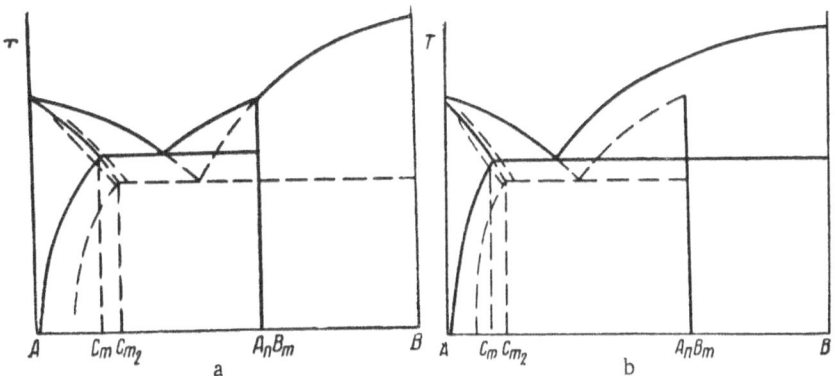

Fig. 58. Two types of double-phase diagrams.

These conditions can be determined for real systems. First, one determines the value of the surface tension at the boundary between the mother phase and the stable and metastable phases and then one determines the probability of the formation of nuclei of a given phase at different degrees of supercooling. The probability of the nucleation of crystallization centers is determined by the ratio between the sizes of the critical nuclei of the different phases for a given degree of supercooling. The second stage is the growth of the nuclei. Nuclei of the metastable and stable phases can grow only if their growth rates are sufficiently high. The growth rate is determined by the flow of atoms to the crystallization front which, in turn, is determined by the concentration, mobility, the gradient of chemical potentials of atoms, and the shape of the growing crystal, and consequently, again, by the surface tension.

The mutual position of the curves representing the variation of free energy with the concentration at a given temperature, even if they are calculated precisely, allows one to determine only the difference between the concentrations and the difference between partial chemical potentials. The other data, such as the diffusion coefficient, density, etc., are known for most alloys.

It is much more difficult to determine the surface tension at phase boundaries and it is impossible to predict the possibility of nucleation of a given phase without knowing the value of the surface tension. We shall discuss the determination of this important thermodynamic value in the following chapter.

Thus, an increase in the degree of supercooling of a system induces the nucleation and growth of crystals of the metastable phase and finally leads to the establishment of metastability, which results in an increase in the volume free energy of the system.

In a great number of binary systems forming intermediate phases the equilibrium can be represented by a double-phase diagram. These diagrams make it possible not only to determine the possible phase composition of the system crystallizing at high degrees of supercooling but also to determine the processes which will occur during heating of the system. In particular, the creation of carbides of different degrees of stability in alloyed steels and the transformation of some carbides into other carbides can be examined from the viewpoint of the creation of systems with different degrees of metastability. In such systems the equilibrium can be represented by a double-phase diagram with respective solid and dotted lines. Investigation of the formation of structure of alloys at a high degree of supercooling, the study of the properties of these alloys, and the search for methods of reaching high degrees of supercooling* are presently in the early stages of development. Thus, many problems remain unsolved, but even now it is clear that knowledge of crystallization of alloys at high degrees of supercooling will provide alloys with new properties useful in technology.

The mere fact that in systems forming intermediate phase one can obtain solid solutions with much higher concentrations than those of solid solutions obtained by quenching indicates the possibility of creating much stronger alloys. Separate crystallization resulting from supercooling is also very promising in terms of creating

---

* It is assumed that these relationships are essentially the result of a given degree of supercooling.

alloys with superior mechanical properties in cases where they cannot be produced because of the formation of a eutectic. It is quite interesting that films of Co–C and Ni–C alloys (containing up to 3% C) turned out to be very ductile while the same alloys in the ordinary state are brittle when the carbon concentration is of the order of a few tenths of one percent. Unfortunately, there have been so few investigations of the properties of alloys obtained at high degrees of supercooling that no generalization is yet possible.

CHAPTER 7

KINETICS OF COALESCENCE
AND DETERMINATION OF THE SURFACE TENSION
AT THE BOUNDARY OF SOLID PHASES

1. Coalescence

In order to use the modern crystallization theory and the data described in the previous chapter for qualitative analysis of the kinetics of phase transformation in alloys one must know the values of the surface tension at the boundaries between the main phases of the alloy.

At present there is no direct method of determining this important thermodynamic value. The indirect experimental methods used are very complicated or give dependable results only for the surface tension at the boundary of the growing nucleus and the surrounding liquid medium in unicomponent systems.

The present author and V. I. Psarev [193, 194] have proposed an indirect method for determining the surface tension at the cementite-austenite, cementite-ferrite, and graphite-austenite boundaries which is based on experimental data on the kinetics of coalescence of the highly dispersed high-carbon component of the Fe−C alloy.

Coalescence is defined as the gradual enlargement of particles resulting from the displacement of the substance composing these particles from small particles to large particles through the surrounding phase.

Coalescence is characteristic of decomposing supersaturated solutions, solid, liquid, and colloidal. In colloidal and ordinary liquid and gaseous systems the small particles also increase in size as the result of the adhesion of particles. This process is called coagulation as opposed to coalescence, where the small particles themselves do not move.

The coagulation of sols in colloidal solutions has been investigated many times. The mathematics of this process was developed by Smolukhovskii [195]. Since the mechanism of coagulation is different in principle from the mechanism of coalescence, we shall not discuss coagulation.

In spite of the fact that coalescence occurs rather often, very few investigations of the process have been made, although there have been investigations which are indirectly related to this process. Most of these have concerned the mechanical properties of steel [196-201] as a function of the size and number of cementite inclusions or the properties of photoemulsions as a function of the size of silver bromide grains [202, 203].

The dependence of the number of particles of globular cementite on the annealing time of carbon steel at 720°C was determined in [198]. The size distribution of grains of globular cementite as a function of the time of isothermal annealing in a similar steel was investigated in [200].

Interesting data on coalescence are found in [204], in which the qualitative thermodynamic theory of coalescence is described and the variation of the average size of $CuAl_2$ precipitated in Al−Cu alloys during coalescence is given. These data show that the Thompson equation can be used to describe the coalescence mechanism in solid phases.

In [204] the dependence of the average size of particles on the time of isothermal annealing is presented in the following form:

$$r \approx \frac{M_3 V}{RT \ln \frac{C_1}{C_2}} t,$$

where C_1 is the concentration of the mother phase at the boundary with the particle of radius r.

On the basis of the laws of thermodynamics one can assume that a binary system containing one component in the form of very small inclusions is unstable from the thermodynamic viewpoint, since it has an excess of free energy because of a longer separation boundary between the phases:

$$F = \sigma S,$$

where σ is the surface tension and S is the total area of the surface.

Since the free energy tends to a minimum, the process occurs in the direction in which the total separation boundary will decrease. This decrease of the length of the boundary can occur as the result of enlargement of particles of the dispersed phase. These general considerations give only the direction of the process.

To explain the mechanism of coalescence one uses the Thompson equation (3.5). This equation shows that the equilibrium concentration of the mother phase I will be greater at the separation boundary with the particles of the dispersed phase II at which the radius of curvature is smallest. This can be explained by the fact that a considerable number of atoms of phase II on the convex surface with a large curvature have a higher energy and therefore pass more often into the surrounding solution.

Usually the dispersed phase consists of grains of different sizes. As a result, there is a difference in concentrations in the areas of the mother solution adjacent to grains of different sizes, which induces diffusional flow of the substance composing the dispersed phase from smaller particles to larger particles.

The flow of substance from one area to another is described by the diffusion equation

$$dq = D\,\frac{C_1 - C_2}{x}\,S dt. \tag{7.1}$$

Here C_1 and C_2 are determined by Eq. (3.5). Equations (3.5) and (7.1) show, at least qualitatively, the physical magnitudes on which the kinetics of coalescence depends. These equations also show that the surface tension at the phase boundary plays a major role in this process.

Quantitative description of the kinetics of coalescence requires an equation which will relate either the number of particles or the average size to the time of isothermal heating.

The most serious attempts at a quantitative description of coalescence were made in [205] and [206]. Both authors studied isothermal coalescence in binary systems in which phase I was a homogeneous solution (solid or liquid) and phase II consisted of spheroidal particles distributed uniformly in the solution.

In a later theoretical study of coalescence [207] the final equation proposed in [206] was derived on the basis of a detailed and more thorough analysis of the coalescence process. The author in [208] checked the validity of this equation for the description of the coalescence of the dispersed phase in alloys and showed that in most cases this equation is not applicable, since it does not satisfy either the initial or final conditions of the process.

The first data on the value of the surface tension at the boundary between the nucleus of the crystallizing phase and the surrounding liquid in a unicomponent system were obtained by Danilov and his students [12].

The following values of surface tension were found for Bi, Sn, and Pb, for which the metastability limits are respectively $\Delta T \approx 30°$, $30°$, and $3°$:

$$\sigma_{Bi} = 20 \text{ erg/cm}^2, \quad \sigma_{Sn} = 17 \text{ erg/cm}^2; \quad \sigma_{Pb} = 3 \text{ erg/cm}^2.$$

These values of σ appear to be too small if one takes into account the fact that the values of surface tension of metals at the liquid-vapor boundary are of the order of hundreds of erg/cm^2. However, comparison of the heats of fusion and evaporation indicates that the value of surface tension at the liquid-crystal boundary should be one-tenth that at the liquid-vapor boundary [114].

Numerous investigations of the structure of different liquids and the determination of crystallization parameters led Danilov to conclude that different types of liquids have a tendency to supercooling. Substances for which x-ray diffraction differs in the liquid and solid states are easily supercooled. In other words, if the stacking of atoms in the liquid and solid states is very different, then at the liquid-solid boundary there is a large interphase surface tension, which increases the work of formation of nuclei.

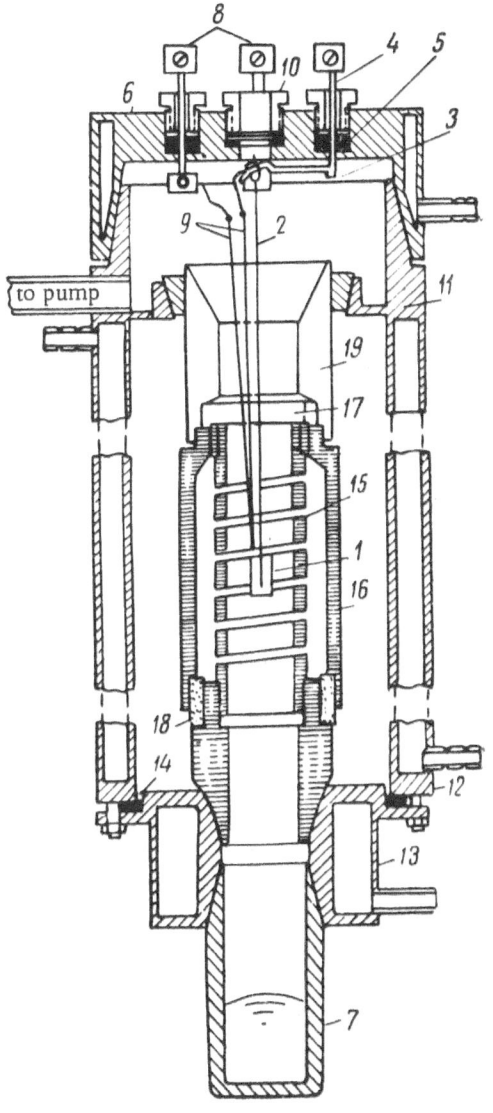

Fig. 59. Vertical vacuum furnace.

Holloman and Turnbull proposed an original method for indirect determination of the surface tension at the nucleus-liquid boundary which is based on the determination of the degree of supercooling. They used this method to determine the surface tension at the nucleus-liquid boundary of many unicomponent systems (see Table 13).

In [209] the same method was used to determine the surface tension at the boundary between solid and liquid iron. Thus:

$$\sigma_{Fe+0.005\%\,Al} = 140-150\ erg/cm^2;\quad \sigma_{Fe+0.8\%\,C} = 120\ erg/cm^2;$$

$$\sigma_{Fe+2\%\,C} = 105\ erg/cm^2;\quad \sigma_{Fe+5\%\,C} = 115\ erg/cm^2.$$

In [211] the authors used the method of calculating the surface tension at the phase boundary proposed in [210] to determine the surface tension at the boundary between austenite and a growing pearlitic colony at a temperature of 650-700°C. For steel containing ~0.003% boron, σ = 10.2-10.4 dyne/cm. For steel containing no boron, σ = 9.7 dyne/cm.

In [212, 140, 41] it was found that the surface tension at the carbide-iron boundary is of the order of 1000 dyne/cm. In [213] a value twice as large was obtained.

Recently [214, 215], the variation of the surface tension at the grain boundaries of crystals of a given metal was determined as a function of their mutual orientation.

There are very few data on the surface tension at boundaries between solid phases.

The surface tension of iron carbide at the boundary with iron was determined by using the Thompson equation, which takes into account the variation of the solubility of the surrounding solution with the change of the radius of the particles of the dispersed phase. The difficulties in determining an infinitely small variation in the concentration make the values of surface tension obtained somewhat doubtful.

2. Experimental Data on the Kinetics of Coalescence

The coalescence of small particles of graphite in eutectic wrought iron containing 1% Si, 0.5% Mn, 0.15% P, and 0.1% S was studied in [216]. The wrought iron was smelted in a Tamman furnace and then poured into a copper chill mold. Samples were cut from the resulting plates, which were 40 × 60 × 6 mm. The samples were quenched and then annealed in a vacuum furnace at 900° C. Those samples with the smallest grains and with graphite inclusions most uniform in size and shape were chosen from these samples.

In [218, 219] the coalescence of carbides was studied in the following steels: 1) pure carbon steel (based on Armco iron) containing 1.7% C; 2) steel containing 2% Co and 1.1% C; 3) steel with 2% Ti and 1.4% C; 4) steel with 2% Mo and 1.6% C; 5) standard ShKh15 steel; 6) standard U8 steel.

All except the standard steels were smelted in argon in a high-frequency furnace. After smelting, the steels were forged into bars 15 × 15 mm and these were annealed. The samples were subjected to isothermal annealing, during which coalescence actually did occur, in a lead bath or in the vacuum furnace.

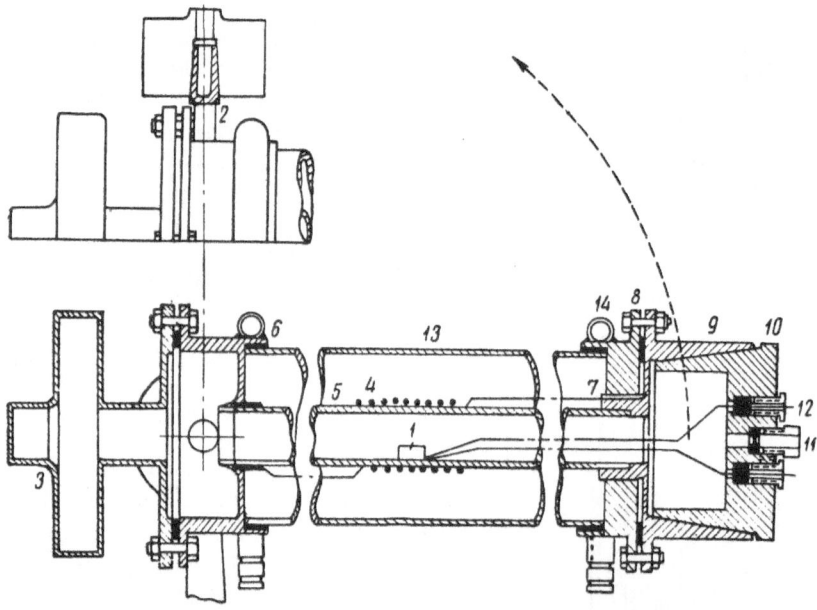

Fig. 60. Horizontal quenching furnace.

In the furnace (Fig. 59) the sample 1 with the wire holder 2 was suspended on the hook 3 which activates the rod 4. The sample is dropped when the hook 3 is turned 100° by the rod 4 through the rubber gasket 5 in the furnace cover 6. The sample falls into a glass filled with mercury (or vacuum oil) 7. The furnace cover 6 has two other holes through which pass contacts 8 insulated with mica and rubber. One end of a Pt-PtRh thermocouple 9 is connected to the contacts 8 and the other is connected to the sample. There is a window 10 between the leads of the thermocouple. The furnace consists of a pipe 80 mm in diameter. A conical end section 11 is welded to the pipe and is closed with a cover 6 at the upper end and a collar 12 at the lower end. A cooling jacket for the furnace is welded to the flanges. The cover is also cooled with water. A cylinder 13 with a conical side, bolted to the collar 12, is cooled with water. The glass containing the mercury is ground to fit the lower part of the conical side of the cylinder and a graphite conductor is fitted to the upper side of the cylinder.

The cylinder 13 is electrically insulated from the body of the furnace by a rubber gasket 14; a textolite bushing insulates the bolts. The graphite heating element 15 is a helix chiseled out of a sleeve of a graphite electrode 50 mm in diameter. The pitch of the helix is 8 mm (the same as the sleeve). The outer diameter is 32 mm and the inner diameter 28 mm. After chiseling the sleeve and the adjustment of the outer diameter, all the ten turns were cut by hand. The lower end of the heater is set into the graphite current carrier and the upper end is fixed in the graphite cylinder 16 with a graphite ring 17. The cylinder 16 is insulated from the lower heat conductor by a ceramic ring 18. The upper end of the cylinder 16 is connected to the body of the furnace by the current carrier 19. This heater raises the temperature to 1700°C under 12 V and 100 A. A single heater can be used for more than 10,000 h at an average temperature of 1000°C. The degree of vacuum and the rate of evacuation depend on the vacuum pumps used. No carbonization of the samples was observed. We also tested molybdenum and tungsten heaters with iron current carriers in this furnace; under this condition the rate of evacuation increases but the resistance of these heaters is somewhat lower than that of the graphite heater, particularly at temperatures above 1200°C. The outer diameter of the furnace is 115 mm and its length 900 mm.

A second type of furnace is shown in Fig. 60. In this furnace the sample 1 and holder are placed in the horizontal position. During quenching the furnace is placed in the vertical position by rotating around the stem 2 through which the furnace is evacuated. As a result of this rotation the sample and holder fall into the vessel 3 containing the quenching liquid. In this furnace a nichrome (or tungsten or molybdenum) spiral heating element 4 is wound on a porcelain (or quartz) tube 5. The ends of the heating element are held with metal rings which

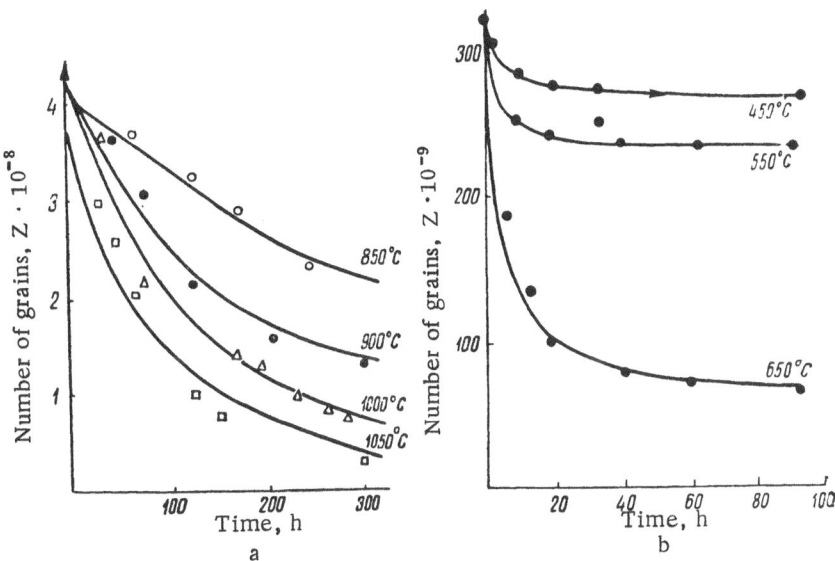

Fig. 61. Coalescence of inclusions. a) Carbon in cast iron
resulting from annealing; b) cementite in U8 steel.

fit into a depression in the end flange 6 and ring 7 insulated from the end flange 8. The ring 7 is connected to the cylinder 9 with copper wires and a ground plate 10 is fitted to the cylinder 9. The plate 10 has a viewport 11 and the leads from the thermocouple 12 pass through it. The bolts holding the cylinder 9 to the flange 8 are insulated with mica plates. Flat vacuum rubber rings are placed between the end flanges, the vessel 3, and the cylinder 9. The end flanges are welded to the pipe 13 and are cooled with water. This requires only one or two windings of a copper tube 14, 8 mm in diameter, which is welded to the surface of the flange. In some cases the outer metal tube is replaced with a quartz tube; then the energy spent heating the sample decreases and the furnace has a higher inertia, particularly if it is coated with asbestos oil. In this way the furnace can be used for long periods of isothermal heating. The rings of the flange are fixed to the quartz tube with a picein or Mendeleev putty. The maximum outside diameter of the furnace (diameter of the vessel) is 85 mm and the length is 1000 mm.

Figure 68(p.110) shows a micrograph of a polished sample of steel heated 15 min at 950°C and quenched in vacuum (mercury).

The sample is first polished, etched (steel is etched in sodium picrate), and then the number of grains per cm² of surface is counted under the microscope (using a grid).

The grains are counted in several areas and the results totaled and averaged.

The number of grains of dispersed phase per unit volume is calculated by the equation

$$z_V = (z_s)^{3/2}.$$

Then the sample is placed in the vacuum furnace, where it is kept at a constant temperature for a given time (vacuum of the order of 10^{-3} mm Hg). After this, a layer 0.5-1 mm thick is ground off the sample and the number of grains again counted. From each original sample a "witness" is cut off to determine the initial size distribution of grains.

When a posteutectoid steel is heated above the temperature A_1, pearlite is precipitated from austenite. The occurrence of this structural component prevents counting of cementite grains which are in the system during isothermal heating above A_1; therefore, the sample is quenched.

Samples of cast iron were annealed at 850, 900, 1000, and 1050°C. The variation of the number of grains with the time of isothermal annealing at different temperatures is shown in Fig. 61. These data indicate that coalescence is very rapid at the beginning (the first 100 h) and then much slower. The figure also shows that the coalescence rate increases very rapidly with increasing temperature.

Even after prolonged annealing (of the order of 300 h) there were no globular graphite inclusions. On the contrary, these inclusions tend to become elongated during annealing. The elongation is essentially along the boundaries between the crystals of austenite. The higher the temperature, the more intense the elongation and branching of inclusions.

The coalescence of carbides follows approximately the same pattern except that the carbides retain their globular shape for a longer time and then become polyhedrons (Fig. 62). Data on the variation of the average radius of cementite grains with the time of isothermal annealing are given in Tables 5-9. The variation of the coalescence rate with the temperature in samples of U8 steel containing the same concentration of cementite as the initial samples is shown in Fig. 61b.

The study of the coalescence of carbides in the steels containing 1.49% C and 2% Ti, and 1.6% C and 2% Mo gives very different results.

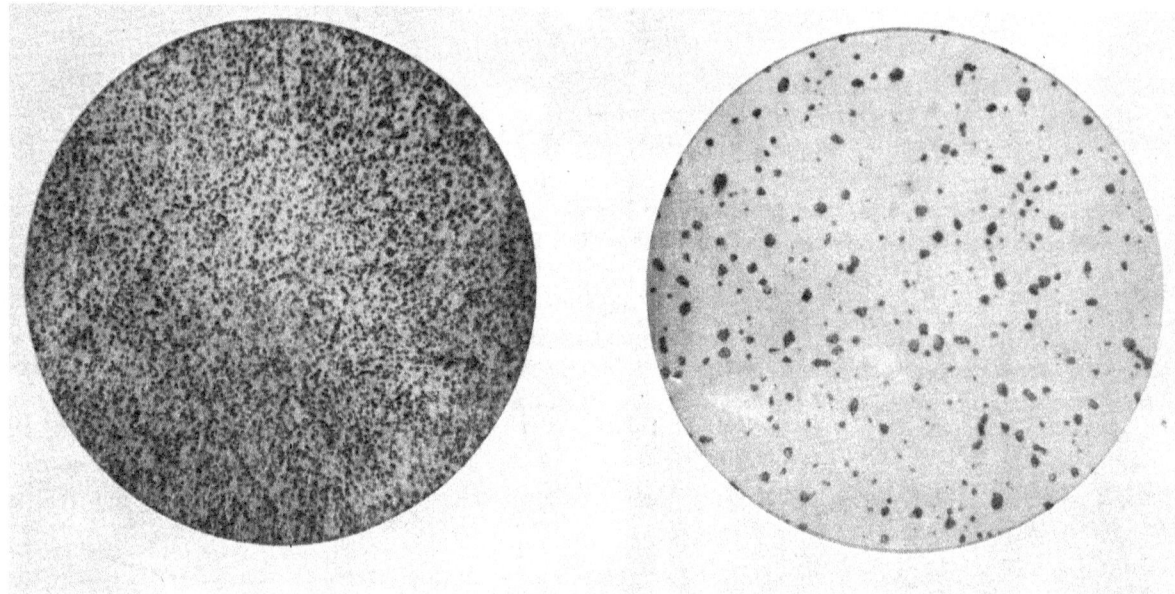

Fig. 62. Coalescence of cementite in posteutectic steel. Left) Initial sample;
right) same sample after 20 h at 850°C. ×700.

In the beginning, the coalescence in these alloys follows the same pattern as in ordinary carbon steel (Fig. 61a). In the following stage crystallites of special carbides begin to separate from the mass of dispersed particles. The shape and characteristic distribution of the special carbides in the matrix of the alloy are very different from those of the rest of the particles. From the moment of the occurrence of these carbides the coalescence is reduced essentially to the "absorption" of cementite particles by the crystallites of special carbides [208, 220].

All this indicates that coalescence proceeds in a somewhat different way when the system contains carbides with different compositions. Numerous investigations have shown that special carbides do not occur during isothermal heating but are present in the mass of dispersed particles in the initial sample. Coalescence only "puts in evidence" the special carbide particles which occur during high-temperature annealing of quenched initial samples. The special carbides are the stable phase, and when they precipitate they decrease the solubility of carbon in the surrounding mother phase. This leads to the diffusion of carbides from the areas of the matrix adjacent to cementite grains to the special carbides, and as the result the cementite grains dissolve and the special carbides grow.

TABLE 5. Variation of the Average Radius of Cementite Grains as a Function of the Time U8 Steel Is Kept at Different Temperatures

T = 623°K			T = 723°K			T = 823°K			T = 923°K		
t, h	$r_{eu} \cdot 10^5$ cm	$r_T \cdot 10^5$ cm	t, h	$r_{eu} \cdot 10^5$ cm	$r_T \cdot 10^5$ cm	t, h	$r_{eu} \cdot 10^5$ cm	$r_T \cdot 10^5$ cm	t, h	$r_{eu} \cdot 10^5$ cm	$r_T \cdot 10^5$ cm
0	4.45	4.45	0	4.4	4.4	0	4.402	4.4	0	4.97	4.27
2	4.54	4.54	1.5	4.5	4.49	1	4.65	4.65	1	4.65	4.65
5.5	4.59	4.62	5	4.58	4.6	4.5	4.71	4.7	4	5.1	5.3
10.5	4.62	4.63	10	4.6	4.62	9.5	4.74	4.78	9	5.67	5.8
20.5	4.65	4.69	20	4.65	4.67	20	4.85	4.85	19	6.25	6.4
40.5	4.7	4.7	35	4.68	4.78	40	4.89	4.89	39	6.8	6.8
60.5	4.72	4.728	55	4.69	4.8	60	4.93	4.9	59	7.15	7
102	4.73	4.73	104	4.7	4.87	100	4.95	4.95	98	7.4	7.3

TABLE 6. Variation of the Average Radius of Cementite Grains in Steel Containing 1.7% C with the Time the Steel Is Kept at Different Temperatures

T = 1023°K			T = 1073°K			T = 1123°K			T = 1143°K			T = 1173°K		
t, h	$r_{eu} \cdot 10^5$ cm	$r_T \cdot 10^5$ cm	t, h	$r_{eu} \cdot 10^5$ cm	$r_T \cdot 10^5$ cm	t, h	$r_{eu} \cdot 10^5$ cm	$r_T \cdot 10^5$ cm	t, h	$r_{eu} \cdot 10^5$ cm	$r_T \cdot 10^5$ cm	t, h	$r_{eu} \cdot 10^5$ cm	$r_T \cdot 10^5$ cm
0	5.75	5.75	0	5.9	5.9	0	7.83	7.83	0	7.24	7.24	0	8.63	8.63
1.75	7.15	7.9	1.75	8.5	10.4	0.5	9.4	9.46	0.25	8.9	8.34	0.5	11.5	11.1
2.75	9.07	9.1	2.75	12	12.4	1.5	12	12.4	0.75	12.8	10.3	1	14.67	13.5
4.75	9.77	11	4.75	17.3	16.3	3	15.6	16.4	1.75	15.63	13.8	2	20.7	18.2
9.75	14.37	15.17	9.75	21.6	21.26	6	21.8	22.7	2.75	18	16.2	5	27.35	29.2
13.25	16.32	17.1	14.75	25.6	24.6	11	22.6	30.3	4.75	19.3	19.8			
18.25	20.1	20.4	19.75	27.2	27	15	26.8	34	7.75	19.9	23.7			
			24.75	28	28.4	30	34.3	38.5	12.75	23.8	27.9			
			30	29	30	26.5	43.1	42.8	17.75	32.6	30			

3. Surface Tension at Cementite-Ferrite, Cementite-Austenite, and Graphite-Austenite Boundaries

In what follows we describe the calculations of the kinetics of the coalescence process in view of determining the order of magnitude of the surface tension at the boundary of the solid phase on the basis of experimental data given in the previous section.

The results of calculation can also be used to determine the kinetics of coalescence necessary for the development of the technology of heat treatment, which is related to this process. The dispersed phases coalesce when the system contains particles of different sizes.

The equilibrium concentration of the mother phase at the boundaries with particles of the dispersed phase depends on their radii and is determined by the well-known Thompson equation. The concentration of the substance composing the dispersed phase in the mother phase is greater at boundaries with small particles than at boundaries with large particles. The concentration gradient leads to dissolution of the small particles and to

TABLE 7. Variation of the Average Radius of Cementite Grains in ShKh15 Steel as a Function of the Time the Steel Is Kept at Different Temperatures

	T = 873°K			T = 953°K			T = 993°K	
t, h	$r_{eu}\cdot 10^5$ cm	$r_T\cdot 10^5$ cm	t, h	$r_{eu}\cdot 10^5$ cm	$r_T\cdot 10^5$ cm	t, h	$r_{eu}\cdot 10^5$ cm	$r_T\cdot 10^5$ cm
0	4.987	4.987	0	4.985	4.985	0	4.983	4.983
3	5.36	5.35	3	5.78	5.78	3.5	5.73	5.9
8	5.64	5.67	8	6.14	6.4	8.83	6.59	6.6
18	5.795	5.78	18	6.91	6.9	18.83	7.36	7.3
38	5.95	5.93	35.86	7.17	7.2	49.16	7.6	7.9
50.67	5.96	5.97	48.5	7.2	7.3			
70	5.98	6	70	7.4	7.46			
130	6.05	6.05	130	7.59	7.57			

TABLE 8. Variation of the Average Radius of Cementite Grains in Steel Containing 1.1% C and 2% Co as a Function of the Time the Steel Is Kept at Different Temperatures

	T = 823°K			T = 873°K			T = 893°K			T = 923°K			T = 953°K	
t, h	$r_{eu}\cdot 10^5$ cm	$r_T\cdot 10^5$ cm	t, h	$r_{eu}\cdot 10^5$ cm	$r_T\cdot 10^5$ cm	t, h	$r_{eu}\cdot 10^5$ cm	$r_T\cdot 10^5$ cm	t, h	$r_{eu}\cdot 10^5$ cm	$r_T\cdot 10^5$ cm	t, h	$r_{eu}\cdot 10^5$ cm	$r_T\cdot 10^5$ cm
0	5.8	5.8	0	5.7	5.7	0	5.9	5.9	0	5.48	5.5	0	6.77	6.78
5	5.83	6	3	6.02	6.09	5	6.3	6.31	2	5.9	5.9	2	7.2	7.15
15	6.04	6.37	8	6.4	6.67	19	7.73	7.2	5	6.37	6.42	3	7.97	7.35
35	7.11	7.08	12.58	7.03	7.09	32	8.62	7.9	10	7.57	7.2	5	7.98	7.64
55.5	8.06	8.06	19.58	7.15	7.6	50	8.67	8.4	20	8.59	8.4	9	9.15	8.2
			32.58	7.4	8.3	69.7	8.99	9	35	9.04	9.5	14	8.8	8.8
			48	7.66	8.87				51.25	11.5	10.5	20	9.22	9.27
			60	8.37	9.02				72	11.15	11.27	35	11.04	10.3
			73	8.93	9.2							45	11.1	11.06
			92.3	9.4	9.6							60	11.9	11.2
												78	11.4	11.4

growth of the large particles, and this process continues until the average size of the particles becomes so large that the concentration at the deformed boundary will be almost the same as that at the plane boundary or until the distance between the particles of the dispersed phase becomes so large that the transfer of the substance from one particle to another will cease for practical purposes.

In terms of coalescence, all the particles of the dispersed phase can be divided into two classes:

1) Large particles ("absorbing" particles). These are the particles which continue growing up to the moment the process practically ceases. Usually, the dispersed phase consists almost exclusively of these particles at the moment the process practically ceases.

2) Small particles (the "absorbed" particles). In the initial stages of coalescence the smallest particles dissolve. The largest particles of this type can even increase somewhat, but since the particles of the first type are present in the system, these particles will still dissolve in the later stages.

Experiments show that the average sizes of particles of the first and second types increase continuously during the process of coalescence.

TABLE 9. Variation of the Average Radius of Cementite Grains in Steel Containing 1.49% C and 2% Ti as a Function of the Time the Steel Is Kept at Different Temperatures

T = 873°K			T = 893°K			T = 923°K			T = 953°K		
t, h	$r_{eu} \cdot 10^5$ cm	$r_T \cdot 10^5$ cm	t, h	$r_{eu} \cdot 10^5$ cm	$r_T \cdot 10^5$ cm	t, h	$r_{eu} \cdot 10^5$ cm	$r_T \cdot 10^5$ cm	t, h	$r_{eu} \cdot 10^5$ cm	$r_T \cdot 10^5$ cm
0	6.05	6.05	0	6.92	7	0	6.16	6.16	0	7.76	7.76
5	6.98	6.98	2	7.2	7.18	5	8.31	8.33	2	8.3	8.32
10	8.16	8.1	5	7.63	7.61	10	9.94	9.8	5	11	9.1
15	8.34	9.5	8	7.7	7.9	20	11.27	11.8	10	11.3	10.2
19	9.6	10.2	13	8.1	8.5	30	12.87	13.1	20	14	12.3
24	12.2	11.2	18	9.74	9.2	45	14.3	14.2	30	14.7	13.8
34	13.2	12.8	23	11.9	10.2	55	14.5	14.8	40	15.12	15.1
			38	13.25	11.4						
			53	13.9	12.8						
			70	14.1	14.09						

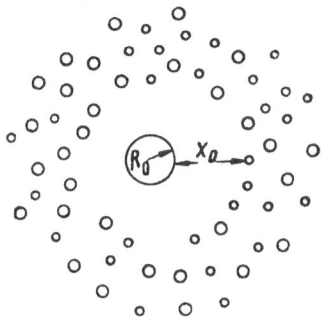

Fig. 63. Model for the calculation of the kinetics of coalescence.

Also, the growth rate of particles of the dispersed phase decreases all the time, and the more rapid the growth, v, the greater the change in the growth rate dv/dt; therefore, we can write:

$$dv = \alpha v\, dt,$$

where α is a constant.

After integration, we obtain:

$$-\ln v = \alpha t + A.$$

When t = 0, v = v_0, and therefore A = $-\ln v_0$.

Also, v = dr/dt, and consequently:

$$dr = v_0 e^{-\alpha t}\, dt.$$

Integration gives us

$$r = -\frac{v_0}{\alpha} e^{\alpha t} + A_1.$$

When t = 0, r = r_0, and consequently,

$$A_1 = r_0 + \frac{v_0}{\alpha}.$$

Substituting the value of A_1, we obtain

$$r = r_0 + \frac{v_0}{\alpha}\left(1 - e^{-\alpha t}\right). \tag{7.2}$$

If we extent $e^{-\alpha t}$ into series and use only the first terms in the series, we obtain

$$r = r_0 + \frac{v_0}{\alpha}\left(1 - \frac{1}{1+\alpha t}\right) = r_0 + \frac{v_0 t}{1+\alpha t}.$$

The values of v_0/α can be determined from (7.2). In fact, when $t \to \infty$, $r_1 = r_m$, and consequently $r_m = r_0 + v_0/\alpha$, and therefore $v_0/\alpha = r_m - r_0$.

Then:

$$r = \frac{r_0 + r_m \alpha t}{1 + \alpha t} = \frac{r_0 + at}{1 + \frac{a}{r_m} t}, \qquad (7.3)$$

where $a = r_m \alpha$.

Let us investigate the growth mechanism of one particle of the first type surrounded with smaller particles among which there are particles of different sizes (Fig. 63). In the initial stages of coalescence the smallest particles dissolve not only next to particles of the first type but also away from them, among particles of the second type. These particles dissolve because some particles of the second type are larger. Thus, a space free of the dispersed phase is created around the growing particle. It can be represented as a hollow sphere through which the substance from the front of the small particles diffuse. The radius of the inner surface of the sphere is at the same time the radius of the growing particle R_1. The outer radius of the sphere is the sum $R_1 + x$, where x is the thickness of the sphere.

The concentration of the mother phase at the outer sphere C_1 depends on the average radius r_1 of the "absorbed" particles; the concentration at the surface of the growing particle C_2 also depends on the radius. These dependences can be described by the approximate Thompson equation:

$$C_1 = C_\sim \left(1 + \frac{2\sigma MV}{RTr_1}\right); \quad C_2 = C_\sim \left(1 + \frac{2\sigma MV}{RTR_1}\right).$$

The amount of the substance diffusing through the hollow sphere and precipitated on the large particle at the time t is determined by the following relationship:

$$q = 4\pi D \int_0^t (C_1 - C_2) \frac{R_1 + x}{x} R_1 dt, \qquad (7.4)$$

where D is the diffusion coefficient and t is time.

By replacing the value of the concentration and representing the distance from the surface of the "absorbing" particle to the front of the "absorbed" particles as $x = bR_0$, where b is a constant determined by the initial conditions, we can rewrite Eq. (7.4) as

$$q = \frac{8\pi D\sigma MVC_\sim (b+1)}{RTb} \int_0^t \frac{R_1 - r_1}{r_1} dt. \qquad (7.5)$$

At the same time, the same amount of substance can be expressed as

$$q = \frac{4\pi}{3V} (R_1^3 - R_0^3), \qquad (7.6)$$

where V is the specific volume of the substance of the growing phase. If we express Eq. (7.5) in differential form and equate it to the time derivative, we obtain

$$K \frac{R_1 - r_1}{r_1} = \frac{4\pi}{V} R_1^2 \frac{dR}{dt},$$

where

$$K = \frac{8\pi D\sigma MVC_\sim (b+1)}{RTb}.$$

The value of dR/dt can be determined by Eq. (7.3), assuming that the size of particles of the first type changes with time according to the law

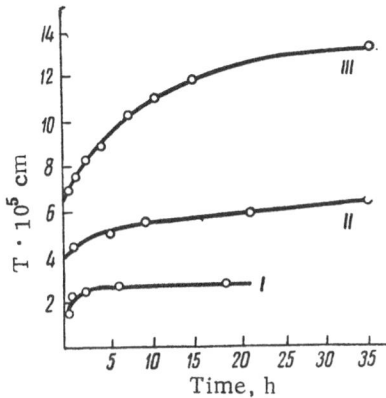

Fig. 64. Coalescence of cementite according to: 1) [200]; II) [198]; III) [218].

$$\frac{dR_1}{dt} = \frac{a\left(1 - \dfrac{R_0}{R_m}\right)}{\left(1 + \dfrac{a}{R_m}t\right)^2}.$$

By substituting dR_1/dt, we obtain

$$K\frac{R_1 - r_1}{r_1} = \frac{4\pi}{V}R_1^2\frac{a\left(1 - \dfrac{R_0}{R_m}\right)}{\left(1 + \dfrac{a}{R_m}t\right)^2}.$$

This relationship must be satisfied regardless of the values of t. When $t \to \infty$ both sides of the equation tend to zero, since $R_1 \to r_1$. Making $t = 0$, we obtain the expression of the surface tension σ at the boundaries of solid phases,

$$\sigma = \frac{aRTR_0^2 r_0 b}{2DMVC_{\sim}(R_0 - r_0)(b+1)}\left(1 - \frac{R_0}{R_m}\right), \qquad (7.7)$$

where R is the universal gas constant; T is the absolute temperature; M the molecular weight of the growing phase; and $C_{\sim}$ the equilibrium concentration at the phase boundary.

Thus, if we have all the necessary experimental data we can determine the surface tension at the boundary of the solid phase. These data must include information on the initial size distribution of grains, from which $R_0 - r_0$ and b are determined. From the curve representing the variation of the average radius of the grain with time one determines a and r_m. The other values can be found in handbooks.

To check the validity of these calculations we first used the experimental results obtained in [200] and in [198]. The most detailed data are given in [200], where all the values necessary to determine σ were obtained. Both publications contain data on the kinetics of the coalescence of globular cementite in carbon steel. The coalescence of cementite at 630°C in pre-eutectoid steel was investigated in [200], while in [198] the author used steel with a different concentration of carbon; isothermal annealing was done at 720°C. Figure 64 shows the dependence r(t) obtained by Eq. (7.3).

The main values used in Eq. (7.7) and the values of σ obtained by using this equation are given in Table 10.

At present it is impossible to compare these data with any other. The applicability of this method of determining the surface tension at the phase boundary can be checked only after a great number of experimental results on different alloys at different temperatures have been obtained. However, the following qualitative criteria favor the method proposed:

1. The order of magnitude of the surface tension obtained in this way is reasonable. This follows from the comparison of the orders of magnitude of the surface tensions at the nucleus-liquid boundary found by Danilov for alloys of different systems.

TABLE 10. The Main Values Entering in Eq. (7.7) and the Values of σ

Steel containing	T, °K	$r_0 \cdot 10^5$ cm	$r_m \cdot 10^5$ cm	$R_0 \cdot 10^5$ cm	$a \cdot 10^9$	b	σ, dyne/cm
0.43% C [3]	903	1.69	2.83	2.05	12.5	1.5	15
0.63% C [7]	993	6.06	23	8.6	2.4	1.93	9
0.82% C [7]	993	6.8	14.8	9.5	3.5	1.93	10
0.97% C [7]	993	6.87	15.6	9.6	4.4	1.7	11
1.16% C [7]	993	7.62	15.6	10.8	3.7	1.67	10

TABLE 11. Main Values for Wrought Iron

T, °K	$r_0 \cdot 10^4$ cm	$r_m \cdot 10^4$	$R_0 \cdot 10^4$	$a \cdot 10^{10}$	b	σ, dyne/cm
1123	3.9	200	5.5	0.9	1.96	73
1173	3.9	200	5.8	2	1.96	71
1273	3.78	200	5.4	3	1.9	54
1323	4.08	200	6	5	2	45

2. The experimental data obtained from different sources give the same value of the surface tension (which drops with increasing temperature), which agrees with the thermodynamic conclusions.

3. The values of surface tension obtained by using the data in [198] indicate that σ on the cementite-ferrite boundary is independent of the concentration of carbon in the steel (above 0.03%). In fact, the increase of the carbon concentration may lead either to the increase of the number of globular cementite particles or their increase in size, but the energy of interaction of atoms does not change during this process.

We used the following data for our calculations by Eq. (7.7):

$$R = 8.3 \cdot 10^7 \text{ erg/deg} \cdot \text{mol}, \quad M = 180, \quad V = 0.13 \text{ cm}^3/\text{g},$$

$$D = 0.02 \exp\left(-\frac{20\,100}{RT}\right), \quad C_\sim = 0.03\% \text{ C}.$$

The experimental and theoretical variations of the average radius of the particles of the dispersed phase with the time and temperature of isothermal heating of different steels and cast irons is shown in Tables 5-9. The values of a and r_m calculated by Eq. (7.3) are given in Tables 11 and 12. Table 12 contains other data on the values of the surface tension at the austenite-cementite, cementite-ferrite, and graphite-austenite boundaries. The arguments in favor of the reasonableness of the order of magnitude obtained can be supplemented by the following:

1. The surface tension of cementite at the austenite boundary is one-third that at the graphite-austenite boundary at the same temperature. This agrees with the phenomena accompanying the solidification of cast iron.

2. In all the alloys investigated the value of the surface tension drops with increasing temperature and the dependence of σ on T is the reciprocal of the dependence of the solubility of the dispersed phase in the mother phase (in the temperature range where there are no phase transformations).

3. Investigations of alloyed steels showed that a relatively small amount of the alloyed element (no more than 2%) does not change the value of the surface tension. However, when carbide-forming elements (Cr, Ti, Mo) are introduced, the surface tension at the carbide-ferrite boundary has a definite tendency to increase, while when a non-carbide-forming element (Co) is added, the surface tension at the same boundary tends to decrease.

4. We can also ascertain that the values of the surface tension obtained by calculation are reasonable by determining the value of the critical nucleus during phase transformation. For instance, if the steel containing 1.2% C is supercooled 50° below the A_{cm} point then the radius of the critical nucleus of cementite is

$$r^* = \frac{2\sigma MV}{RT \ln \dfrac{C}{C_\sim}} = \frac{2 \cdot 12 \cdot 180 \cdot 0.13}{8.3 \cdot 10^7 \cdot 1223 \cdot \ln \dfrac{1.2}{1.13}} \approx 8.9 \cdot 10^{-8} \text{ cm}.$$

4. Surface Tension at the Austenite-Ferrite Boundary

The value of the surface tension at the cementite-austenite boundary obtained in the preceding section makes it possible to calculate the surface tension at the austenite-ferrite boundary. For this calculation we used the main relationship concerning the growth of pearlitic colonies [221].

It is well known that in the growth of pearlitic colonies the growth rates of ferrite and cementite plates are equal.

The growth rate of each of these plates, respectively, u_1 and u_2, can be represented by the following expressions:

$$u_1 = \frac{Dk}{\rho_1}\left(\frac{\Delta_1}{r_1} - \frac{a_1}{r_1^2}\right), \quad u_2 = \frac{Dk}{\rho_2}\left(\frac{\Delta_2}{r_2} - \frac{a_2}{r_2^2}\right),$$

where k is the shape coefficient; ρ_1 and ρ_2 are respectively the densities of ferrite and cementite; Δ_1 and Δ_2 are the respective degrees of supersaturation; r_1 and r_2 are the radii of curvature at the separation boundary and are equal to half the thickness of the respective plates

$$a_1 = \frac{2\sigma_1 M_1 C_{\sim 1}}{RT\rho_1} \text{ and } a_2 = \frac{2\sigma_2 M_2 C_{\sim 2}}{RT\rho_2},$$

where σ_1 is the surface tension at the ferrite-austenite boundary; σ_2 is the surface tension at the cementite-austenite boundary; M_1 and M_2 are the molecular weights of ferrite and cementite; R is the universal gas constant; and T is the temperature.

In the case of eutectoid decomposition of austenite and cementite $u_1 = u_2$, and consequently:

$$\frac{\rho_1 r_1^2}{\rho_2 r_2} = \frac{\Delta_1 r_1 - a_1}{\Delta_2 r_2 - a_2}.$$

We shall represent supersaturation as:

$$\Delta_1 = C_{\sim 1} - C_1, \quad \Delta_2 = C_{\sim 2} - C_2,$$

where $C_{\sim 1}$ and $C_{\sim 2}$ are the equilibrium concentrations of ferrite and cementite in austenite; and C_1 and C_2 are the concentrations of ferrite and cementite in supersaturated austenite.

Let us write Eq. (7.8) in the form:

$$\frac{\rho_1 r_1^2}{\rho_2 r_2^2} = \frac{r_1 C_{\sim 1} - (a_1 - C_1 r_1)}{r_2 C_{\sim 2} - (a_2 - C_2 r_2)}.$$

Since $C_{\sim 1}/C_{\sim 2}$ is nothing else but the ratio between the amounts of ferrite and cementite in the eutectoid, the ratio between the volumes of these plates is determined by the ratio of their thicknesses (their other parameters being identical), we have

$$\frac{\rho_1 r_1^2}{\rho_2 r_2^2} = \frac{r_1 C_{\sim 1}}{r_2 C_{\sim 2}},$$

and consequently

$$\frac{r_1 C_{\sim 1}}{r_2 C_{\sim 2}} = \frac{C_1 r_1 - a_1}{C_2 r_2 - a_2}.$$

As the result of the law of conservation of mass, we have

$$\frac{C_{\sim 1}}{C_{\sim 2}} = \frac{C_1}{C_2}, \quad \frac{r_1 C_{\sim 1}}{r_2 C_{\sim 2}} = \frac{a_1}{a_2} = \frac{\sigma_1 M_1 \rho_2 C_{\sim 1}}{\sigma_2 M_2 \rho_2 C_{\sim 2}},$$

and

$$\sigma_1 = \sigma_2 \frac{M_2 \rho_1 r_1}{M_1 \rho_2 r_2}.$$

If we use the data on the surface tension at the cementite-austenite boundary taken from Table 12, i.e., $\sigma_2 \approx 25$ erg/cm^2, and take into account the fact that in pearlite $r_1/r_2 \approx 7$, then $\sigma_{\text{fer-aus}} = (25 \cdot 180 \cdot 7)/56 = 550$ erg/cm^2, and this corresponds to the value of the surface tension at the ferrite-austenite boundary obtained in [223] and is about one-third the much too high value obtained in [41].

TABLE 12

Steel	T, °K	$r_0 \cdot 10^5$ cm	$K_0 \cdot 10^5$ cm	$z_W \cdot 10^5$ cm	$a \cdot 10^9$	b	σ, dyne/cm
1.1% C, 2% Co	823	5.8	7.1	14.0	0.2	1.5	17
	873	5.7	7.4	11.5	0.8	1.5	15
	893	5.9	7.5	12	0.46	1.5	7
	923	5.5	7.2	15	0.9	1.5	6
	953	6.78	8.9	13.8	1.07	1.5	6
1.7% C	1023	5.75	7	42	4.2	1.94	28
	1073	5.9	7.3	38	10	1.94	25
	1123	7.83	9.8	65	10.7	1.94	23
	1143	7.24	8.7	39	15	1.94	19
	1173	8.63	11	120	15	1.94	15
1.49% C, 2% Ti	873	6.05	7.3	30	0.94	1.22	51
	893	7	7.8	50	0.4	1.22	35
	923	6.16	7.5	18.5	2.2	1.25	29
	973	7.76	9	2.8	1.08	1.29	12
U8	623	4.45	4.6	4.742	3	1.32	152
	723	4.4	4.7	4.9	2	1.32	48
	823	4.4	4.7	4.9	6	1.32	29
	923	4.27	5	7.54	3.2	1.32	15
ShKh15	873	4.987	5.5	3.2	2.8	1.44	27
	953	4.985	5.6	7.65	3	1.44	13
	993	4.983	5.4	8.4	2.5	1.44	9

5. Approximate Equations for Determining the Values of Surface Tension

In this section we give data obtained by calculating the surface tension at the phase boundaries by using the known equations to determine the values of the critical nucleus in the case of crystallization in unicomponent systems or during the decomposition of supersaturated solutions [222]:

$$r^* = \frac{2\sigma T_S M}{L(T_S - T)\rho},\tag{7.8}$$

where σ is the surface tension at the phase boundary; ρ is the density; T is the melting point; $T_S - T$ is the degree of supercooling; and L is the heat of transformation.

If r^* is expressed as $r^* = nr_0$, where r_0 is the minimum radius of the nucleus of the precipitating substance, equal to half of the lattice constant $a/2$, then Eq. (7.8) can be written in the form

$$n\frac{T_S - T}{T_S} = \frac{2\sigma M}{\rho L r_0}.$$

The right side of this equation is a constant, and therefore the product $n(T_S - T)/T_S$ remains constant regardless of the value of T.

The maximum value of $(T_S - T)/T_S \to 1$ when $T \to 0$; however, the minimum value of n, which cannot be smaller than one, must correspond to the maximum value of relative supercooling. Thus, for all T we have

$$\frac{2\sigma M}{\rho L r_0} = 1,$$

and consequently

$$\sigma = \frac{1}{2}\frac{L\rho}{M}r_0.\tag{7.9}$$

TABLE 13. Surface Tension at the Nucleus-Liquid Boundary

Substance	L, erg/g	ρ, g/cm^3	$\frac{a}{2} \cdot 10^8$ cm	σ_{exp}, dyne/cm	σ_{cal}, dyne/cm	Literature source
Al	$3.7 \cdot 10^9$	2.70	2.02	93.0	98.90	226
Sn	$5.6 \cdot 10^8$	7.30	2.90	59.0	61.60	225
Fe	$2.6 \cdot 10^9$	7.86	1.43	201.0	146.0	226
Ni	$2.9 \cdot 10^9$	8.90	1.70	255.0	232.0	226
Bi	$5.6 \cdot 10^8$	9.80	2.30	54.0	56.0	224
Co	$2.5 \cdot 10^9$	8.90	1.80	284.0	200.0	226
Pb	$2.2 \cdot 10^8$	11.30	2.56	33.0	32.30	224
Sb	$1.6 \cdot 10^9$	6.69	2.20	101.0	116.80	226
Ge	$4.0 \cdot 10^9$	5.36	2.80	181.0	240.0	226
Ag	$1 \cdot 10^9$	10.50	2.00	126.0	105.0	226
Au	$6 \cdot 10^8$	19.30	2.00	139.0	115.80	226
Pt	$9.4 \cdot 10^8$	21.40	2.00	240.0	200.0	227
Ga	$7.68 \cdot 10^8$	5.90	2.30	56.0	53.13	224
Pd	$1.52 \cdot 10^8$	11.50	1.90	209.0	174.0	226
Water	$3.2 \cdot 10^9$	1.00	2.20	32.1	35.20	220
Hg	$1.2 \cdot 10^8$	14.20	2.30	28.1	20.40	224
Na	$9.3 \cdot 10^8$	0.97	2.15	20.0	19.40	226

Table 13 gives the experimental values of the surface tension at the nucleus-liquid boundary obtained by different authors and calculated by Eq. (7.9). All the other values entering Eq. (7.9) are shown in the same table.

The agreement between the theoretical and experimental data is quite satisfactory.

Similar reasoning can be used to determine the surface tension at the boundary of phases when the substance precipitates from the supersaturated solution.

In fact, the radius of the critical nucleus of the precipitating phase is

$$r^* = \frac{2\sigma_{12}MV}{RT \ln \dfrac{C}{C_\sim}} , \qquad (7.10)$$

where σ_{12} is the surface tension at the boundary between the precipitating phase and the mother solution; M is the molecular weight of this phase; V is its specific volume; R is the universal gas constant; T is the transformation temperature; C is the concentration of the mother phase; and $C_\sim$ is the equilibrium concentration.

If, again, we replace r^* with nr_0, then Eq. (7.10) can be rewritten as

$$n \ln \frac{C}{C_\sim} = \frac{2\sigma_{12}MV}{RTr_0} . \qquad (7.11)$$

The right side of this equation is constant at a given temperature T; the minimum value of n, i.e., n = 1, should correspond to the maximum value of $\ln C/C_\sim$. The value of $\ln C/C_\sim$ is maximum when the concentration of the mother phase C is maximum. For phase diagrams of the cigar type this maximum concentration is 100%. For phase diagrams of systems with limited solubility this concentration is C_m.

Then, the equation determining the surface tension acquires the form:

$$\sigma_{12} = \frac{RTr_0}{2\rho M} \ln \frac{C_m}{C_\sim} . \qquad (7.12)$$

Experimental and theoretical results obtained with (7.12) are given in Tables 13 and 14.

TABLE 14. Experimental and Theoretical Values of σ

Boundary	T, °K	ρ, g/cm^3	$a \cdot 10^8$ cm	$r_0 \cdot 10^8$ cm	C_∞, %	σ exp, dyne/cm	σ cal, dyne/cm	Remarks
Cementite–ferrite	993 903 823	7.6	4.5	2.3	0.029 0.021 0.012	8.0 13.0 29.0	11.5 12.0 16.8	C_{max} = 0.03%
Cementite–austenite	1023 1073 1123 1173	7.6	4.5	2.3	0.85 1.05 1.15 1.30	28.0 25.3 19(21)* 15.0	32.1 28.3 23.4 20.1	C_{max}= 2%
Graphite–austenite	1123 1173 1273 1323	2.23	2.4	1.2	1.0 1.20 1.50 1.70	73.0 71(90)* 54(70)* 45(50)*	70.8 58.0 39.4 30.2	C_{max} = 2%

* Calculated by the equation given in [207].

Although these semiempirical relationships need further checking, the experimental data available at present are satisfactorily generalized by Eqs. (7.10) and (7.12). Apparently, an even better agreement between the experimental and theoretical data can be obtained by using the coefficient indicating the difference in the structures of the mother phase and the substance crystallizing from it and using Eq. (4.7).

CHAPTER 8

NONEQUILIBRIUM STRUCTURE
OF GRAPHITE FORMING ALLOYS

Various types of steels and cast irons are the most widely used alloys, and consequently most investigations in the physics of metals concern steels. Cast irons are used extensively, but they are usually used for machine parts which are much less critical than those made of steel. However, cast irons have a number of useful properties, among which are outstanding casting quality, high wear resistance, low cost, etc.

Many years of collaboration between engineers and scientists have produced a "superstrong" cast iron which is characterized by the presence of globular graphite. The mechanical properties of cast iron have been improved to such an extent that the number of machine parts which can be made of cast iron increases every year.

A review of the extensive literature [229-235] shows that at the present time there is no complete agreement on the mechanism of formation and growth of graphite nuclei during the solidification of cast iron under different cooling conditions or on the transformation of the metastable system (containing cementite) into the stable system (containing a high-carbon phase — graphite). This latter problem is a particular case of the general problem concerning the creation of metastable and stable systems and the transformation of one into the other.

The formation of the structure of Fe–C alloys should conform to the general laws of crystallization, although it is necessary to take into account the singularities of the components which affect the nucleation and the growth of different phases in this system.

We shall not give here a critique of the different viewpoints concerning the nucleation and growth of phase components, but we shall analyze this process from the viewpoint of the thermodynamics and kinetics of phase transformations, taking into account the known singularities of this system [236, 237].

1. Formation and Growth of Graphite Nuclei

Most metal physicists agree that phase diagrams with double lines should be used for the study of Fe–C alloys. Such diagrams make it possible to explain a great number of facts which are important from the practical viewpoint. And there is another reason to justify the use of double-phase diagrams: the components of these systems form intermediate phases with very little difference between the minimum free energies. Depending on the level of these minima and their mutual positions with respect to the concentration coordinate, different degrees of supercooling are necessary for crystallization of the unstable phase.

The double-phase diagram is a geometric locus of points which determine the equilibrium conditions of the iron solution with each of the high-carbon phases. Since the mother phase of the Fe–C alloys (liquid, austenite, or ferrite) cannot be simultaneously in equilibrium with the very different phases (cementite and graphite) under the same conditions, a phase diagram with double lines becomes a necessity.* The cementite is unstable only when the system contains graphite separated from the cementite by a layer of the mother phase. The different solubilities of the high-carbon components in the mother phase (represented by the solid and dashed lines on the phase diagram) are responsible for the instability of cementite.

On the basis of these assumptions we shall attempt to explain why gray cast iron crystallizes when the degree of supercooling is low and white cast iron crystallizes when the degree of supercooling is high. This can be explained either by the fact that when the degree of supercooling is high graphite centers have no time to form or by the fact that cementite centers grow much more rapidly than graphite centers.

* Or one can use two-phase diagrams, one for Fe–Fe_3C and one for Fe–C.

According to Gibbs, the work of formation of a nucleus is determined by the relationship:

$$F = \frac{1}{3}\,\sigma S^*,$$

where σ is the surface tension and S^* is the area of the separation surface. The value of the surface tension can be determined from the critical radius of the nucleus

$$r^* = \frac{2\sigma MV}{RT \ln \dfrac{C}{C_\sim}}\,.$$

For a given degree of supersaturation of cast iron, the ratio between the radii of the critical nuclei of cementite and graphite are expressed as

$$\frac{r^*_{gr}}{r_{cem}} = \frac{\sigma_{gr}M_{gr}V_{gr}T_{cem}}{\sigma_{cem}M_{cem}V_{cem}T_{gr}},$$

where σ_{gr} is the surface tension of the graphite-mother phase boundary and σ_{cem} is the surface tension at the cementite-mother phase boundary. Calculation of surface tension showed that

$$\frac{\sigma_{gr}}{\sigma_{cem}} \approx 4.$$

Then, if we take the temperature into account, we have

$$\frac{r^*_{gr}}{r_{cem}} = \frac{4 \cdot 12 \cdot 0.44}{180 \cdot 0.13} \approx 0.9,$$

and

$$\frac{F_{gr}}{F_{cem}} = \frac{\sigma_{gr}S^*_{gr}}{\sigma_{cem}S^*_{cem}} = \frac{\sigma_{gr}\,(r^*_{gr})^2}{\sigma_{cem}(r^*_{cem})^2} \approx 3.2.$$

Thus, the work of formation of the cementite nucleus is one-third of the work of formation of the graphite nucleus.[*] If we take into account the fact that under ordinary conditions the degree of supersaturation for the formation of graphite is somewhat higher, we may assume that the graphite nucleus always has time to form in technical cast iron.

Let us now examine the growth rates of graphite and cementite nuclei. The growth rate of a spherical nucleus is given by the equation:

$$v = DV\left(\frac{\Delta}{r} - \frac{a}{r^2}\right).$$

If we express the size of the nucleus as $r = r^* n$, where $r^* = \alpha/\Delta$, we obtain

$$v = \frac{D\Delta^2 RT}{2\sigma MC_\sim}\left(\frac{n-1}{n^2}\right).$$

The ratio of the rates of growth of graphite and cementite centers of the same size is

$$\frac{v_{gr}}{v_{cem}} = \frac{\Delta^2_{gr}\sigma_{cem}C_{\sim cem}M_{cem}}{\Delta^2_{cem}\sigma_{gr}C_{\sim gr}M_{gr}} \approx \frac{1}{4}\,.$$

Thus, for the same size of the nucleus and for the same degree of supersaturation the growth rate of cementite nuclei is much greater than that of graphite nuclei.

---

[*] For the sake of brevity, we shall not examine the variation of crystallization with the composition of cast iron because the initial stages of crystallization, i.e., the formation and growth of centers up to the point where they begin to interact, follows the same mechanism in all alloys.

During the manufacture of white cast iron ledeburite is formed. It is well known that the growth rate of a eutectic colony is much higher than the growth rate of the crystals of each of the phases composing the eutectic. Therefore, cementite nuclei become larger than graphite nuclei even during the early stages of solidification of cast iron. Then the cementite reacts with austenite, forming ledeburite colonies whose growth rate is even higher, thus inhibiting the growth of graphite centers.

If the degree of supercooling is low, the relative positions of the dashed and solid lines indicating the supersaturation of the liquid with respect to graphite and cementite, then the difference in the growth rates is reduced to a minimum, and in some cases the graphite nucleus grows even faster than the cementite nucleus.

The calculation given above is only approximate, since there are no precise data on the surface tension at the boundary between the high-carbon component and the liquid. However, there is no reason to doubt that not far from the eutectic line and at lower temperatures the ratio of the surface tensions used for the calculation is very different from the real value. This assumption is confirmed by the fact that the mutual positions of dashed and solid lines on the phase diagram are almost the same for the solid and liquid states.

Let us now examine the process of precipitation of the high-carbon phase during cooling of austenite in posteutectoid steel.

The solubility of carbon in austenite decreases as the austenite cools. At the austenite-cementite boundary the solubility of carbon in austenite is given by the line ES; at the austenite-graphite boundary the solubility of carbon in austenite is given by the line E'S'. When posteutectoid steel is cooled below these lines a graphite or cementite nucleus is created in the austenite. Thus, when an alloy containing 1.5% C is supercooled to 900°C the critical radius of the graphite nucleus is

$$r_{gr}^* = \frac{2 \cdot 80 \cdot 12 \cdot 0.44}{8.3 \cdot 10^7 \cdot 1200 \cdot \ln \dfrac{1.5}{1.2}} \approx 5 \cdot 10^{-8} \text{cm},$$

and the critical radius of the cementite nucleus is

$$r_{ce\overline{\overline{m}}}^* = \frac{2 \cdot 20 \cdot 180 \cdot 0.13}{8.3 \cdot 10^7 \cdot 1200 \cdot \ln \dfrac{1.5}{1.3}} \approx 8 \cdot 10^{-\varepsilon} \text{ cm}.$$

The work of formation of the nuclei are

$$F_{gr} = \frac{1}{3} \sigma (4\pi r^{*2}) \approx 5.1 \cdot 10^{-13} \text{erg},$$

$$F_{ce\overline{\overline{m}}} 5 \cdot 10^{-13} \text{ erg}.$$

Thus, when austenite in posteutectoid steel is supercooled the growth rate of graphite nuclei may be the same as that of the cementite nuclei.* Calculations similar to those made for the liquid state indicate that the growth rate of the cementite nucleus is about four times higher than that of the graphite nucleus. However, the growth of a graphite nucleus in a solid solution is much more difficult than in a liquid. The great difference in the specific volumes of graphite and austenite results in the fact that the growth of a graphite nucleus is accompanied by plastic deformation as the result of which the iron atoms move away from the graphite crystallization front. It is well known that iron atoms move essentially as the result either of the motion of dislocations or as the result of exchanging places with vacancies. Therefore, the growth rate of graphite grains is determined by the rate of motion of iron atoms under the pressure of the growing graphite nucleus. This assertion is confirmed by comparison of the theoretical and experimental data on the growth rate of the graphite nucleus.

According to our conclusions in Chapter 5, the radius of a grain in the later stages of growth is given by the equation:

$$r = \sqrt{2 \cdot D \cdot V \cdot \Delta \cdot t} + r_0.$$

---

* This concerns spontaneous nucleation.

TABLE 15. Theoretical and Calculated Values of Surface Tension Obtained by I. N. Bogachev ($D = 7.5 \cdot 10^{-3}$ g/cm^3)

T = 1000°C D = 6.6 · 10⁻⁷ cm²/sec			T = 900°C D = 2.7 · 10⁻⁷ cm²/sec			T = 850°C D = 1.9 · 10⁻⁷ cm²/sec		
t, h	r_{exp}, mm	r_{cal}, mm	t, h	r_{exp}, mm	r_{cal}, mm	t, h	r_{exp}, mm	r_{cal}, mm
0.5	0.027	0.093	1	0.016	0.085	5	0.018	0.16
1	0.048	0.13	3	0.054	0.14	10	0.027	0.23
1.5	0.063	0.16	5	0.072	0.18	20	0.063	0.33

The comparison of the theoretical and experimental results obtained in [232] is shown in Table 15. This table indicates that the theoretical growth rate differs from the experimental growth rate by more than one order of magnitude. Therefore, although graphitization centers form during cooling of austenite, their growth is inhibited to such an extent that the cementite nuclei forming at a later stage "outgrow" the graphite nuclei quite easily and stop their growth.

What happens to the graphite nuclei which formed and grew to a certain size? At a high cooling rate nuclei of subcritical size can exist among the crystallization centers; these nuclei are heterophase fluctuations which did not have time to grow to supercritical size (unstable at the crystallization temperature) and to dissolve during the short period of time in which the metal is cooled.

These numerous graphite centers of different sizes are finally surrounded by a phase consisting of austenite and cementite, or ferrite and cementite, or martensite and supersaturated austenite.

Let us examine the behavior of these graphite formations under different conditions.

Depending on its size and the type of environment, such graphite centers are either stable and continue to grow or are unstable and dissolve.

The solubility depends not only on the nature of the phases in contact but also on their degree of dispersity. This assertion has been confirmed by numerous experiments and is described very well by the Thompson equation.

Since the solubility depends on the size of the nuclei, not all the graphite nuclei will be stable and continue to grow in the presence of relatively large cementite crystallites. If the concentration of austenite at the boundary with the graphite nucleus of radius r is $C_1 = C_{\infty}(1 + 2\sigma M/\rho RTr)$ and the concentration at the boundary with large cementite crystallites is C_2, then graphite centers can grow only when $C_1 < C_2$. If we assume that the surface tension of graphite at the austenite boundary at t = 1100°K is about 100 dynes/cm and $r = 10^{-7}$ cm, we have

$$\frac{C_1}{C_2} = \frac{1.1}{1.2}\left(1 + \frac{2 \cdot 100 \cdot 12}{2.3 \cdot 83 \cdot 10^7 \cdot 10^7 \cdot 1100}\right) \approx 1.03.$$

Consequently, nuclei with a radius of 10^{-7} cm are unstable in the presence of cementite, and therefore it is the graphite and not the cementite which will dissolve. The stability of these nuclei increases somewhat with increasing temperature (the surface tension decreases) and with increasing degrees of dispersity of cementite.

Even this relatively rough calculation shows that the sizes of graphite nuclei can explain some phenomena accompanying graphitization. More precise data on the size of stable graphite nuclei can be obtained when the distance between the dashed and solid lines of the phase diagram is known. Also, it is necessary to take into account the fact that the radius of curvature of the cementite–austenite boundary is often negative (particularly in ledeburite).

All these discussions lead us to assume that not all the graphite nuclei can be transformed into stable graphitization centers during annealing of white cast iron. Fluctuational nucleation of graphite centers in the austenite of white cast iron and steel during isothermal annealing appears to be improbable.

It is improbable because the austenite is in equilibrium with cementite, which damps significant deviations of the system from the equilibrium composition, and the size of the nucleus which is stable in the presence of cementite is too large for spontaneous nucleation.

When the cooling rate of cast iron is high, pearlite (sorbite, troostite), martensite, supersaturated austenite, or bainite can form when the cast iron passes through the A_1 point.

Under favorable conditions graphite inclusions of subcritical size can be stable in the presence of dispersed cementite and increase to larger sizes.

The most favorable surrounding for the growth of graphite centers is martensite-austenite. Graphite inclusions surrounded with supercooled austenite or martensite begin to grow at a high rate as soon as carbon atoms are capable of moving as the result of diffusion.

In this light, the role of preliminary quenching is reduced to ensuring that the small graphite inclusions present in the system are surrounded with a medium which provides the best conditions for growth of these inclusions during secondary heating of the alloy.

Thus, depending on the cooling rate, graphite nuclei of very different sizes are formed in solidifying cast iron. The size distribution of the centers is typical. The average size of graphite centers is displaced toward smaller values with increasing degrees of supercooling. Under industrial conditions solidification of white cast iron ingots is always accompanied by the formation of a certain amount of free graphite; these graphite inclusions are apparently the centers which occur at the same time as (or before) the nucleation of cementite nuclei.

The amount of free graphite present in white cast iron cooled at a very high rate (as determined by chemical analysis) is sufficient for subsequent graphitization of white cast iron on the already existing graphite centers during annealing.

According to various authors [238, 239], the minimum amount of graphite in white cast iron is 0.005%. This amount corresponds to a volume of $1.7 \cdot 10^{-4}$ cm^3 of graphite per cm^3 of cast iron. The number and the size of graphite inclusions which can be formed from such an amount of graphite are given below:

Radius of the inclusion, cm. . . . . 10^{-7} 10^{-6} 10^{-5} 10^{-4}
Number of inclusions per cm^3 of cast iron . $4 \cdot 10^{16}$ $4 \cdot 10^{13}$ $4 \cdot 10^{10}$ $4 \cdot 10^{7}$

The maximum number of carbon inclusions resulting from annealing of wrought iron does not exceed 10^7-10^8 per cm^3.

These results indicate that the number of nuclei in white cast iron is much greater than the number of carbon inclusions due to annealing formed during graphitization of white cast iron.

In steels the graphitization centers occur when the steel is cooled below the E'S' line.

The solubility of carbon in austenite* decreases with decreasing temperature. The S'E' and SE lines indicate respectively the solubility limits in the case of graphite-austenite and cementite-austenite equilibrium. Cooling of posteutectoid steel or cast iron leads to supersaturation of austenite. Depending on the degree of supersaturation, graphite or cementite nuclei form in austenite. In this case the nuclei are formed and grow in the solid matrix, and consequently considerable volume changes occur and are accompanied by plastic deformation of the mother phase.

The formation and growth of graphite nuclei in Co-C and Ni-C alloys are quite similar. In these alloys the carbide is formed at a considerably higher cooling rate than in Fe-C alloys because the Co_3C and Ni_3C carbides are much less stable than Fe_3C. Consequently, these carbides cannot for practical purposes inhibit the crystallization of graphite at relatively low cooling rates because of the large distance between the dashed and solid lines on the phase diagram.

Thus, any factor which in some way favors the increase in the distance between the lines of stable and metastable equilibrium favors the precipitation of graphite. The factors which slow down the growth of the carbide crystallization centers act in the same way.

_____
* What follows refers to ferrite.

2. Surface Crystallization of Graphite and the Role of Defects

It has been known for a long time that graphite precipitates in places where different types of voids accumulate. These voids may be cracks, porosities, gas bubbles, interdendritic porosities, etc.

Also, a phenomenon called "migration of carbon on the surface" is often observed during solidification of cast iron. This phenomenon consists in the fact that the surface of cast iron becomes covered with graphite powder after the cast iron solidifies.

This phenomenon became of special interest as the result of the investigation of vacuum heat treatment of posteutectoid steels in which graphite was observed to form on the surface of all samples [240-242].

In [244, 245, 236] it was explained by the singular role of some surfaces formed during the crystallization of graphite. It was found that graphite forms on the surfaces of samples only in the case of cooling of a supersaturated solution of carbon in iron (in the solid or liquid state).

Let us analyze whether the precipitation of graphite on the surface of a sample is possible from the viewpoint of the theory of phase transformation. If there is a ready surface (this can be the surface of a section or the inner surface of some void) then nuclei can form not only within the supersaturated solution but also on the surface. The formation of nuclei on the surface has two specific singularities: 1) The nuclei of the new phase can be two-dimensional, which facilitates nucleation considerably; and 2) the formation of a nucleus requires the work of formation of two surfaces of separation (nucleus-medium and nucleus-mother phase) during the liquidation of the surface of separation between the mother phase and the medium. Here, the medium means either the surrounding gas or vacuum. The work of formation of a rectangular two-dimensional nucleus is determined by the equation

$$F_1 = KS\,(\sigma_{12} + \sigma_1 - \sigma_2),$$

where K is the shape coefficient, close to 1/3; σ_{12} is the surface tension at the graphite-mother phase boundary; σ_1 is the surface tension of graphite; σ_2 is the surface tension of the mother phase at the boundary with the medium; and S is the area of the surface of contact between graphite and the mother phase (we can neglect the side surfaces for a two-dimensional nucleus) (Fig. 65).

The formation of an identical nucleus within the matrix requires the work (for simplicity we shall use a two-dimensional nucleus):

$$F_2 = K\sigma_{12} \cdot 2S,$$

and, consequently, the favorable condition for the formation of a new phase on the surface is $F_2 \geq F_1$, or

$$\sigma_{12} \geqslant (\sigma_1 - \sigma_2). \tag{8.1}$$

This indicates that only a nucleus of the phase with a surface tension lower than the surface tension of the initial can form on the surface, and the difference between these surface tensions must be smaller than the surface tension at the boundary between two phases.

The first condition limits the possibility of surface crystallization in different alloys considerably. However, this condition is satisfied for graphite because the basal plane of graphite has a very small surface tension and σ_2 increases with increasing degrees of supersaturation. If the austenite in posteutectoid steel is cooled below the E'S' line then a degree of supersaturation satisfying condition (8.1) can always be reached. As the result, the graphite nucleus occurring on the surface has much more favorable growing conditions than the nucleus formed within the sample. In fact, in order for the nucleus within the sample to grow, the atoms of the solid solution must be displaced, while on the surface the nucleus grows quite freely.

The number of graphite nuclei and their size are determined by the degree of supersaturation reached. This is particularly important in Fe−C alloys because with increasing degrees of supercooling cementite nuclei can form in the matrix and also in its surface layer. The cementite prevents the growth of graphite nuclei because it absorbs the carbon precipitated from the supersaturated solution.

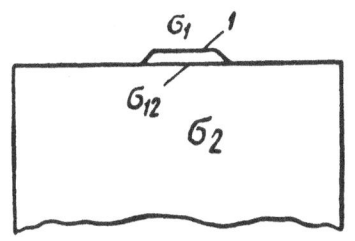

Fig. 65. Formation of a two-dimensional nucleus of graphite due to precipitation. 1) Two-dimensional graphite nucleus; 2) austenite.

Clearly, surface graphitization depends also on the factors determining graphitization within the sample, with the difference that the rate of formation and growth of graphite nuclei are much greater on the surface than within the supersaturated solution.

Up to now we have considered the creation and growth of nuclei in the case of supersaturated solutions which are the result of supercooling. However, surface graphitization can also occur during isothermal heating if the system contains sufficiently large graphite nuclei and cementite.

In other words, the growth of carbon inclusions resulting from annealing can occur on the surface as well as within the sample, only at a much higher rate. Therefore, one usually distinguishes two kinds of surface graphite: carbon due to precipitation and carbon due to annealing. Carbon due to precipitation is formed directly by precipitation from supersaturated solutions, while carbon due to annealing results from the growth of ready-formed nuclei which result from the dissolution of cementite in white cast iron or from the graphitization of steel. Experimental data on surface graphitization are in complete agreement with these assertions, which follow from the crystallization theory.

3. Method of Investigation

The nature, the mechanism, and the kinetics of the formation of surface carbon in Fe–C, Co–C, and Ni–C alloys capable of being graphitized were investigated in [246–252].

A method of vacuum etching described in [254] makes it possible to determine the structure of heated alloys. This method can be modified to fit various operations carried out in vacuum.

This method of investigation by heat treatment in vacuum is of particular importance for the clarification of the mechanism of phase transformation. In this method metals can be heated and cooled (up to quenching) at different rates in vacuum. Under these conditions the surfaces of polished samples or samples cast in vacuum are not oxidized and can be examined under the microscope without preliminary treatment.

At the present time, the study of the structure and properties of alloys during heat treatment in vacuum is a separate branch of metal science. Special apparatus have been constructed which make it possible to observe not only the variation of the structure of the sample during heat treatment but also the variation of the mechanical properties.

The different apparatus and different methods used in vacuum metallography are described in the monograph by Lozinskii [243].

Aside from vacuum heat treatment, it is also interesting to study the surface structure of alloys heated in different gases. With only minor changes, the same type of apparatus can be used for this purpose.

Although there are industrial apparatus which can be used for such investigations, we describe furnaces used in investigating the graphitization of alloys. These furnaces are unique in that they are easy to build in the laboratory and are very easy to operate.

Two of these furnaces were described in Chapter 7. The diagram of the third (smelting furnace) is shown in Fig. 66.

The body of the furnace 1 is made of a seamless pipe to which is welded a bottom 2 cut from a solid piece of metal. The heater 3 is a graphite tube which is connected to the graphite conductors 4, 5. At the top they are connected by a slit graphite cone and at the bottom by a conical contact. A magnesite heat insulation cylinder 6 is placed between the heater and the conductor 4.

A cylinder 7, insulated from the body of the furnace by a rubber sheet and from the bolts by ebonite washers, is bolted to the base of the furnace. The conductor 5 is connected to the cylinder 7 by a conical contact. A thermocouple passes through the glass 8 in the cylinder. This glass can be replaced with a glass filled with quenching liquid (mercury). A device for dropping the sample 10 into the crucible, the mold 11 and the view port 12 are attached to the cover 9.

The device for dropping the sample consists of a rotating perforated disc. The samples are placed in the perforations, and as the disc rotates they fall onto the stationary plate under the disc. The body of the furnace, the cover, and the lower cylinder are cooled with running water.

As the furnace rotates around the conical pin 13 the molten metal flows along the magnesite tube with a funnel mouth 14 (fixed with the ring 15) into the mold. The material and the shape of the mold are chosen to suit the purpose of the investigation.

The furnace is heated by the current from a step-down transformer. A rubber ring is used as a gasket between the body of the furnace and the cover.

RVN-20 low vacuum pumps are used first, and then TsVL-100 diffusion pumps, to create a vacuum of the order of 10^{-3} mm Hg. The temperature is of the order of 1800°C.

When a polished sample of posteutectoid steel is heated in vacuum to the austenization temperature and cooled to room temperature then there are gray spots on the surfaces of some samples (shown in Fig. 67). These spots are some modification of the precipitated carbon. The type and shape of these spots depend on the composition of the alloy and the heat treatment used. The internal structure of the spots also depends on the temperature at which precipitation occurs.

The surfaces of the samples were studied by different methods after the various heat treatments.

The graphite layers were separated from the surfaces of the samples by two methods: by electrolytic dissolution of the metal in Popov electrolyte (current density 0.03 A/mm², voltage of 1.5-2 V, temperature of the electrolyte 10-15°C) or by mechanical removal, using gelatin. In the first case the surface of the sample was cut into squares before the graphite film was detached. After thorough washing in distilled water the pieces of the film were removed with a net and examined under the electron microscope under a magnification of × 3000-6000.

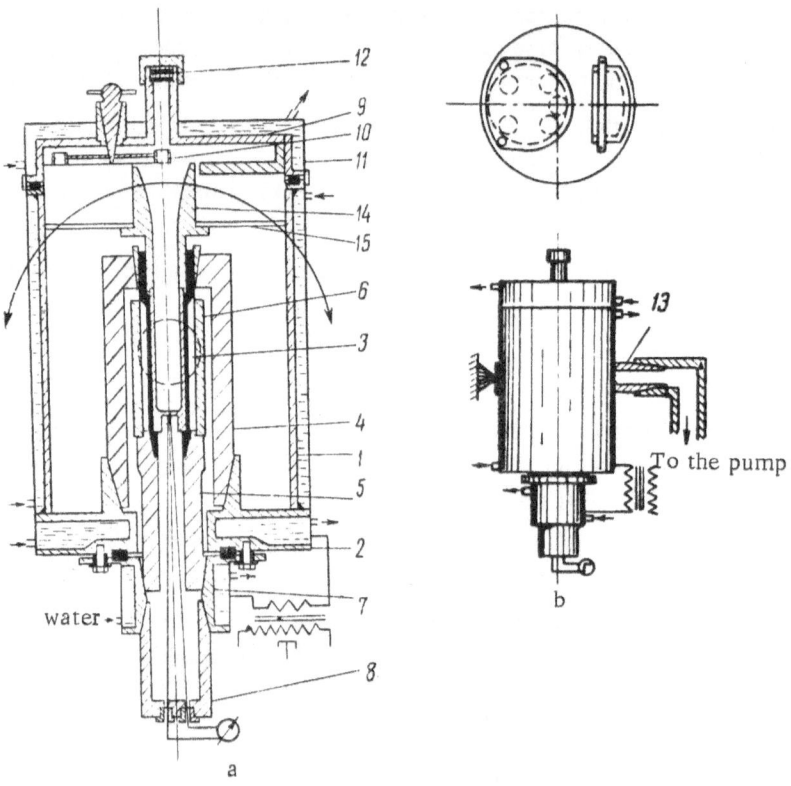

Fig. 66. Vacuum smelting furnace. a) Cross section; b) side view.

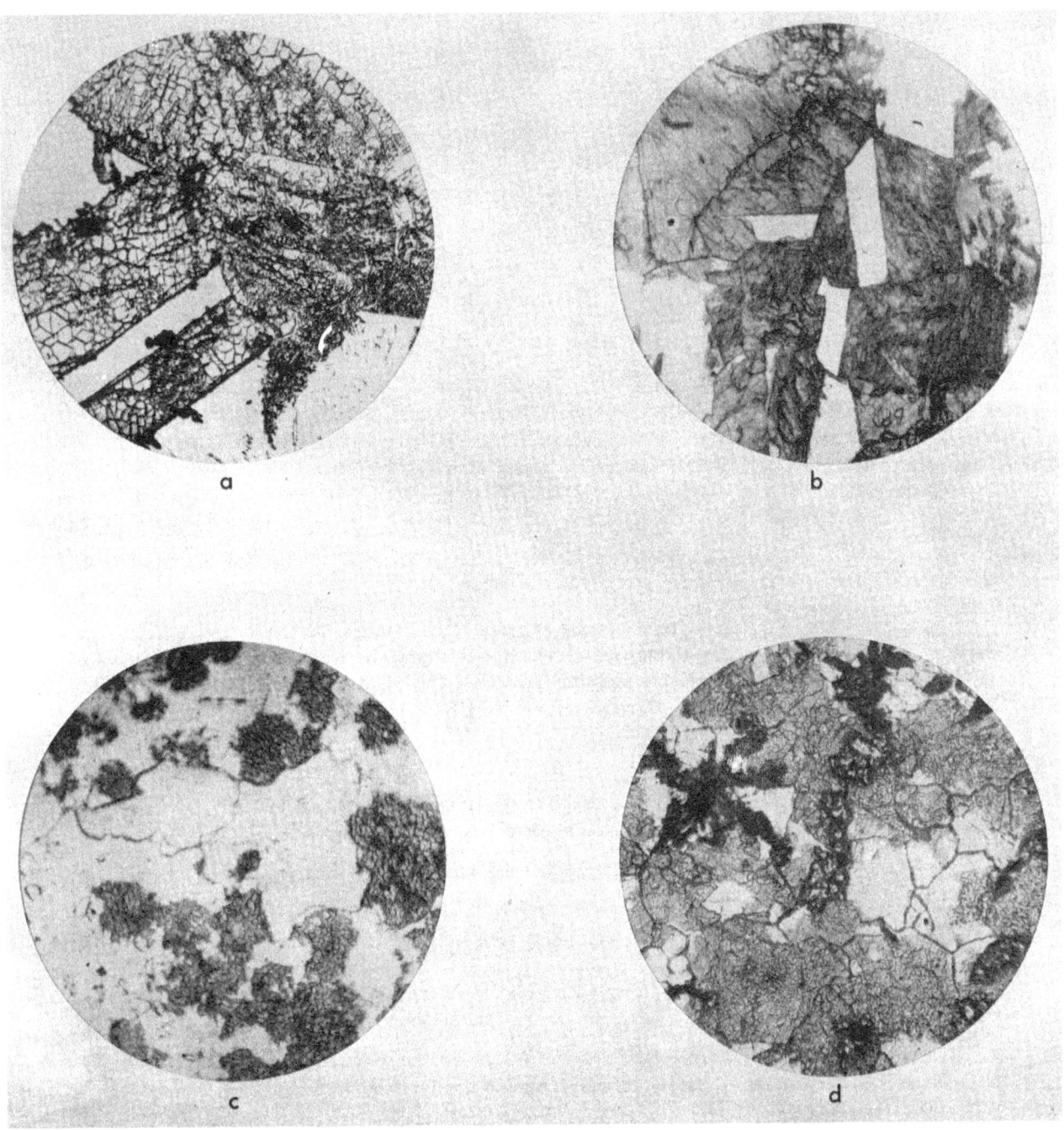

Fig. 67. Graphite precipitated on the surface. a, b) Posteutectoid steel; c) low-carbon steel
containing less than 0.1% C; d) wrought iron. × 200.

In some cases the structures of the precipitated graphite layers were compared with those of graphite layers obtained by evaporation and condensation of carbon in vacuum, using the usual method. The graphite precipitates can be separated from the surfaces of Ni-C and Co-C alloys much easier either electrolytically or mechanically than the layers of condensed graphite.

The graphite films precipitated on the surface of Fe-C alloys were usually removed mechanically, although in some cases it was possible to separate them electrolytically.

Fig. 68. Quenched in vacuum.

The precipitated graphite layers are strong and stable, withstanding washing and the effect of the electron beam. Their thickness changes with the type of heat treatment and the type of crystal face upon which they are precipitated. The electron micrographs were made with the maximum current passing through a collimating lens placed in the projection lens.

4. Carbon Due to Precipitation

Graphite formed on the surface of graphitizable alloys as the result of heat treatment in the solid state is shown in Fig. 67. This graphite was first observed on the surface of posteutectoid steel. Let us examine the principal data obtained as the result of investigations of posteutectoid steel.

1. The graphite precipitates on the surfaces of the samples only during cooling. This result was obtained by quenching steel in vacuum.

In all cases where the sample is heated to high temperatures and then quenched no graphite is formed on the surface. The micrograph in Fig. 68 shows the surface of a sample of posteutectoid steel quenched in mercury. To see whether the graphite drops off during quenching in vacuum, the samples covered with graphite were quenched in vacuum. However, no graphite dropped off the samples.

Prolonged isothermal heating up to 100 h at any temperature followed by quenching consistently produces a clear surface. Graphite precipitates on the surface only when the sample is cooled at temperatures between A_{c_m} and A_1.

2. The graphite precipitates on the surface of the sample only when posteutectoid steel is cooled in the temperature range between the E'S' and S'K' lines.

This was determined in the following way. Samples of different carbon steels containing from 0.4 to 1.7% C were heated and then cooled at different rates. No graphite formed on the surfaces of eutectoid or preeutectoid steels. In all cases the graphite precipitated on samples of posteutectoid steel. Then, different samples of posteutectoid steel were heated to different temperatures and cooled slowly to a temperature somewhat below the E'S' line and then quenched.

These experiments showed that graphite precipitates only on samples cooled 30-50° below the E'S' line. The graphite precipitates during the whole cooling period down to the KS line.

TABLE 16

Experiment No.	Concentration of elements, %				
	C	Mn	Si	S	P
1	0.04	0.04	0.08	No greater than 0.02	No greater than 0.02
2	0.06	0.04	0.08	0.02	0.02
3	0.07	0.04	0.11	0.03	0.03
4	0.1	0.14	0.11	0.03	0.03
5	0.17	0.04	0.08	0.02	0.02
6	0.40	0.04	0.08	0.02	0.02
7	0.48	0.71	0.21	0.05	0.05

3. Graphite is formed on the surface of austenite in posteutectoid steel because of the formation and growth of graphite centers.

This result was obtained in experiments previously described in detail. Often, the graphite centers are formed at the boundaries of austenite grains or at the twinning boundaries of austenite crystallites. Then these centers grow toward the middle of the surface of the grains. In some cases the graphite centers are formed in the central part of the surface of the austenite grains and then grow toward the boundaries.

Sometimes there are so many graphite centers that they cover the surface completely.

4. Not all the crystallites on the surface of the sample are equally covered with graphite. The number of austenite crystallites covered with graphite increases with the amount of carbon in the sample and with a decreasing cooling rate.

In samples containing a large amount of carbon (1.7%) almost all the austenite crystallites become covered with graphite in the case of a slow cooling rate.

At a high cooling rate there is some selectivity in the graphite covering. The austenite crystallites which are not covered with graphite are sprinkled with cementite needles (Fig. 67b).

In samples containing a small amount of carbon (U10 and U12 steels) graphite precipitates on the surface only when the cooling rate is low (1-5°/min), the number of austenite crystallites covered with graphite is small, and the covering itself is very thin.

5. If the polished surface of the sample is etched with a 3-5% solution of nitric acid before it is placed in the vacuum furnace, then the rate of precipitation of graphite is much higher. This is particularly noticeable on samples containing low amounts of carbon.

Apparently, graphite centers form more easily on the etched surface.

6. When the samples covered with graphite are kept at a temperature above the E'S' line the graphite film dissolves in austenite.

Samples covered with graphite were reheated in vacuum up to a temperature above the E'S' line and then quenched. The graphite covering then disappeared.

Repeated heating and subsequent slow cooling leads to precipitation of graphite in areas of the surface which are not necessarily the same as those where they were precipitated originally.

7. If posteutectoid steel is heated to a temperature below the A_{cm} point then whether the graphite precipitates on subsequent cooling depends on the shape, distribution, and amount of cementite remaining in the sample at that temperature. If there are rather large areas free of cementite then graphite will precipitate in these areas. When cementite (globular) is uniformly distributed and separated by small areas of austenite, graphite does not precipitate on cooling.

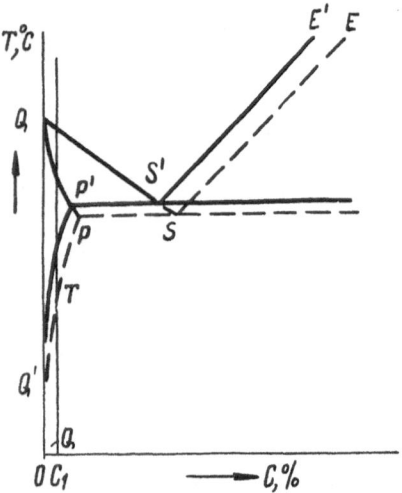

Fig. 69. Section of the Fe—C phase
diagram.

Thus, these experiments confirm the assumptions made in the preceding section concerning the mechanism of the formation of graphite on different ready made surfaces. Graphite precipitates as the result of its decreased solubility in surface layers of austenite during a gradual decrease in temperature.

It is interesting to investigate the processes occurring on the surfaces of samples of low-carbon steel during heat treatment in vacuum (graphitization of ferrite) from this particular point of view.

The chemical composition of the steels investigated is given in Table 16.

Samples of these steels were polished and placed in the vacuum furnace in which the samples could be heated and cooled at different rates up to quenching. The quenching liquid was mercury.

The samples were heated to 680-900°C and cooled at the rate of 40, 20, 10, 5, 2, and 1°/min and then quenched from 850°C.

Metallographic studies of the surfaces showed that when the cooling rate is low (2°/min or less) graphite in the form of gray layers similar to precipitated graphite is formed on the surfaces of samples containing 0.1% C. Figure 67c shows the structure of the surface of the sample covered with precipitated graphite.

When the cooling rate and quenching rate are high, microscopic examination reveals no graphite precipitated either on these samples or the samples containing 0.2% or more carbon.

These results can be explained as follows. The solubility of carbon in ferrite is lower in the case of ferrite-graphite equilibrium than in the case of ferrite-cementite equilibrium. Therefore, the line of stable equilibrium on the phase diagram of Fe—C alloys (Fig. 69) must also be drawn for the ferrite-graphite equilibrium – Q'P'.

Graphitization centers can occur when the alloy with a concentration C_1 is cooled to temperature T. The most probable place for the occurrence of these centers is the surface of the sample. On further cooling of the alloy the graphitization centers can increase in size as the result of the decrease in the solubility of carbon in ferrite. The centers on the surface will grow at the highest rate, since in this case the iron matrix does not prevent their growth and the surface diffusion accelerates the supply of carbon to the crystallization front.

It is obvious, therefore, that at slow cooling rates the graphite nuclei on the surface have time to grow to visible sizes.

When the cooling rate is high the degree of supersaturation can be high enough that the formation of cementite centers becomes probable. The growth rate of these centers is very high, since they require much less carbon. As a result, the cementite centers eliminate the supersaturation and prevent the graphite centers from growing to visible size.

When the carbon content is high, the formation of graphite nuclei of visible size on the surface is inhibited, since the large amount of cementite (contained in the pearlite) surrounding the ferrite grains eliminates the supersaturation which would result from cooling.

The occurrence of graphite on the surface of austenite grains in which the eutectoid transformation occurs is not very probable for the following reasons. When austenite is cooled below the P'S' line the formation of graphite centers on the surface is possible, but they can grow to visible size only when the cooling rate is very low – much lower than the rate used in our experiments. This is possible, first of all, because a low degree of supercooling leads to the occurrence of cementite which, together with the ferrite, forms a pearlite colony which inhibits the growth of graphite. Secondly, the precipitation of carbon from austenite results in the rebuilding of the γ-lattice into the α-lattice. Therefore, a relatively "thick" layer of ferrite is formed under the graphite,

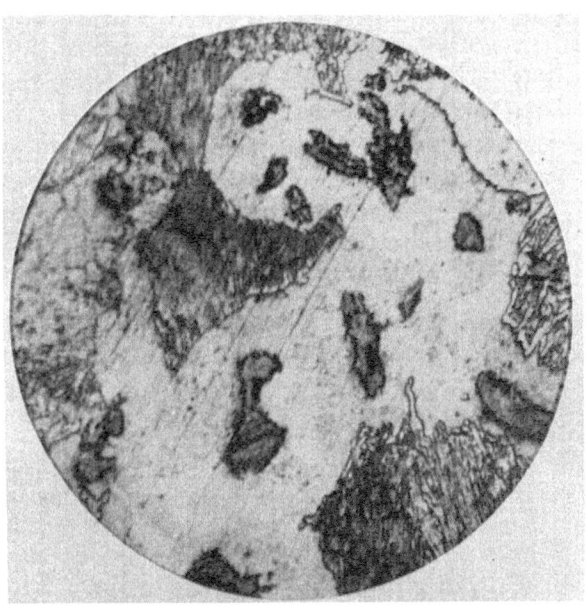

Fig. 70. Structure beneath the precipitated graphite.
× 800.

and this layer insulates the surface of the graphite nuclei from austenite. The ferrite layer is in thermodynamic equilibrium with graphite, and therefore it does not have the necessary concentration gradient between the graphite-ferrite and austenite-ferrite boundaries.

Thus, the precipitation of carbon from ferrite can, to a certain extent, explain the cause of graphitization of low-carbon steels.

In fact, at low cooling rates the inner surfaces of some microdefects (porosities, cracks, etc.) become covered with graphite and many then play the role of graphitization nuclei.

We made a similar investigation with wrought iron.

The micrographs (Fig. 67d) show the structure of a sample of wrought iron (2.3% C, 1.1% Si, 0.5% Mn) after 30 min at 1000°C in the vacuum furnace followed by cooling at the rate of 30°/min.

The boundaries of austenite grains are clearly visible on the surface. Most of the austenite crystallites are covered with a gray layer of graphite. Some of the grains remain white and are not covered with graphite.

The round black inclusions are carbon resulting from annealing. The thickness of the graphite layer is uneven and in some places there are large graphite crystallites. If the surface of such samples is polished and etched one finds that the areas of the metal matrix which were covered with graphite consist of ferrite. The areas on which graphite does not precipitate usually have a pearlitic structure. Figure 70 shows a micrograph of such a sample after polishing and etching in a 3% solution of nitric acid.

Even this brief description of the precipitation process of graphite on the surface of wrought iron indicates that the graphite is formed on the surface because of the decrease in the solubility of carbon in γ-iron during the cooling of wrought iron.

In fact, when the sample is cooled from 1150 to 730°C the solubility of carbon in γ-iron decreases from 2 to 0.9%. During cooling, the major portion of this carbon precipitates on carbon inclusions in wrought iron resulting from annealing. Carbon precipitated from austenite situated close to the surface may form new crystallization centers directly on the surface of the sample.

Graphite precipitated on the surface of wrought iron remains almost undissolved on repeated heating and the metal matrix is saturated with carbon as the result of the carbon inclusions due to annealing. This phenomenon can be explained by the fact that carbon inclusions due to annealing are not equiaxial and have many microprotrusions with a very small radius of curvature. Consequently, the solubility of carbon in austenite at the boundary between austenite and the carbon inclusions due to annealing is higher than the solubility at the boundary between austenite and carbon, which is almost plane.

Cyclic heating or prolonged heating at high temperatures increases the amount of graphite on the surface because the carbon in the carbon inclusion due to annealing is "pumped" to the surface. As the thickness of the graphite layer increases, the recrystallization process begins, which decreases the free energy of the forming layer, since a great number of previously formed crystallization centers are transformed into a relatively small number of large graphite crystals.

As we have said, the presence of cementite inhibits the precipitation of graphite on the surface. This is particularly true in the case of white cast iron. The heating and cooling of polished samples of white cast iron of eutectic composition does not usually lead to precipitation of graphite. When the carbon concentration is

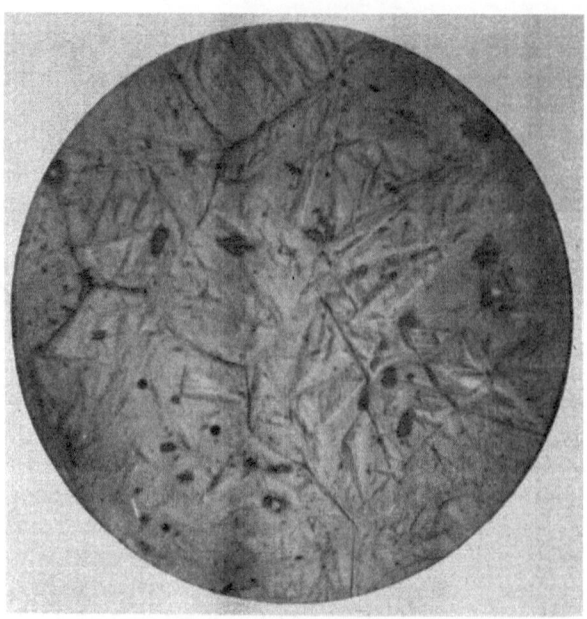

Fig. 71. Graphite formed on the surface of a Ni—C alloy. a) Slow cooling rate; b) an annealed quenched sample. × 500.

low and cementite forms a layer between relatively large austenite grains, small amounts of carbon precipitate at a certain distance from the cementite. For this to occur, the sample must be heated to high temperature and kept there until the cementite inside the austenite grains completely disappears, and the cooling rate must be such that supersaturation is not prevented by the growing cementite.

Role of Quenching and Precipitation of Graphite in Co—C and Ni—C Alloys

The amount of graphite precipitated on the surface from austenite decreases sharply with increasing cooling rates of posteutectoid steel.

However, in steels containing a large amount of carbon it is apparently impossible to stop the formation of carbon nuclei completely. This is confirmed by the fact that subsequent surface graphitization resulting from annealing of quenched steel is much more rapid in samples not polished after quenching. If samples quenched in vacuum are polished (apparently this removes the graphite nuclei) the formation of carbon on the surface due to annealing begins much later.

The high rate of formation of graphite nuclei due to precipitation is also confirmed by the fact that in posteutectoid steels containing a large amount of carbon (of the order of 1.7%) or silicon (~ 0.5%) it is very difficult to prevent the formation of graphite on the surface even at cooling rates at which the martensite transformation occurs (Fig. 71).

At the same time, tempering of quenched steel does not lead to precipitation of graphite on the surface (only carbon due to annealing is formed as the result of prolonged annealing). This is apparently the result of the fact that the cementite nuclei are formed in the martensite and austenite during tempering and during their growth these nuclei absorb all the carbon precipitated from the martensite and supersaturated austenite.

The data on the formation of graphite in Ni—C and Co—C alloys confirm this assumption.

In Ni—C and Co—C alloys no metastable phase of the martensite type is formed in the solid state during quenching, and the solid solution must be highly supersaturated for the carbide to form. Therefore, graphitization (i.e., precipitation of the stable phase) proceeds in such alloys without the difficulties occurring in Fe—C alloys.

Sections of the phase diagrams for Co—C and Ni—C alloys are shown in Fig. 45 (p. 72). On the Ni—C phase diagram the C_m point corresponds to approximately 1300° C and 0.65%C; at room temperature the limit solubility of carbon drops to 0.05-1%. In the Co—C system the limit solubility of carbon in cobalt is about 0.8% at 1300° C and drops to 0.1% at room temperature.

Thus, in this case the solubility of carbon in the solid solution (ferrite) also decreases with temperature. The precipitation in this case is accompanied by a considerable increase in volume. Therefore, graphite similar to the carbon precipitated in Fe—C alloys must precipitate on the surface of such alloys during the cooling of the solid solution. The micrograph of the surface of a Ni—C alloy containing 0.23% C heated to 1100°C and cooled in vacuum shows that almost the entire surface of the sample is covered with graphite (Fig. 72a). The same situation occurs in Co—C alloys. When these alloys are quenched from high temperature in vacuum, graphite does not precipitate on the surface in the same way as in steel samples.

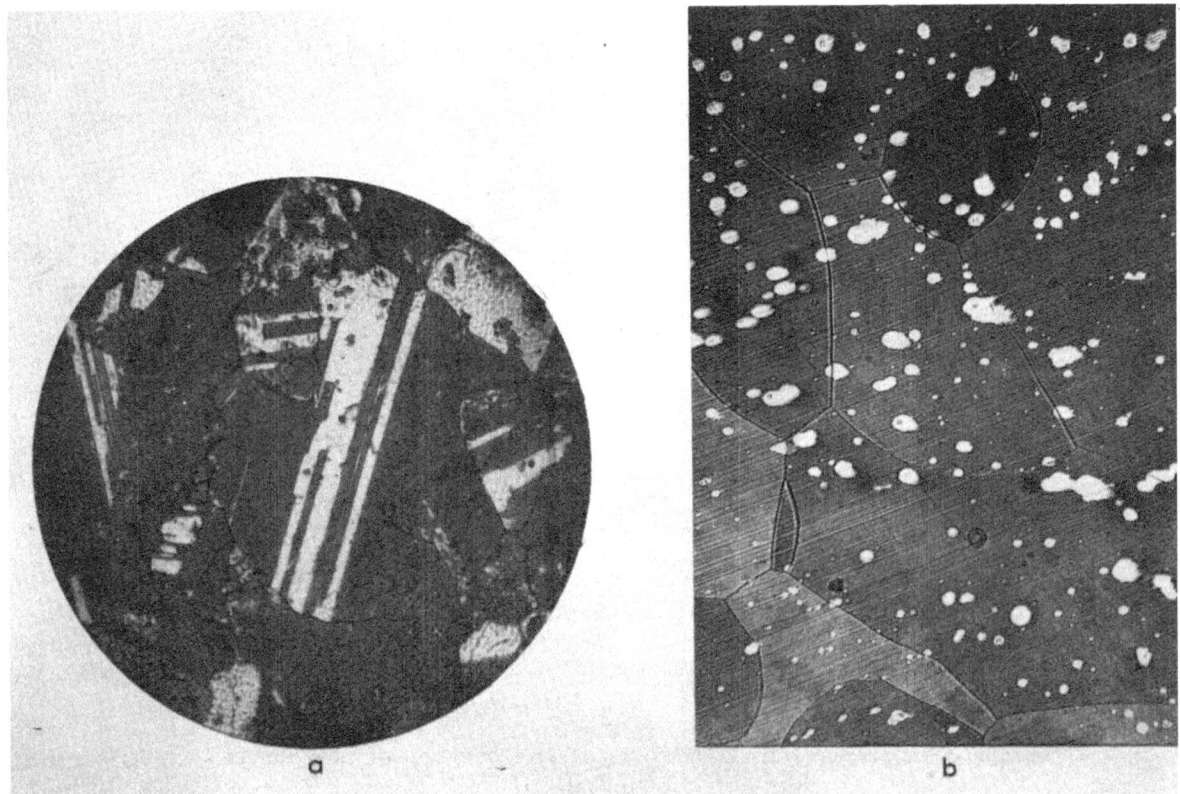

Fig. 72. Graphite precipitated on the surface of a Ni–C alloy. a) Slow cooling rate; b) an annealed quenched sample. × 200.

However, unlike the tempering of Fe–C alloys, the tempering of quenched samples of Ni–C and Co–C alloys leads to the precipitation of graphite on the surface. Figure 72b shows the micrograph of a quenched sample tempered at 500° C. One can see that the surface is covered with graphite.

All this can easily be explained by the fact that crystallization on the surface also results from the formation and growth of graphite nuclei. When the supersaturation is very high, which occurs during quenching, the rate of formation of nuclei during tempering is much higher than during relatively slow cooling of a sample heated to 1100°C.

In the absence of the metastable phase (cementite in alloys) graphite can precipitate on the surface during tempering of quenched alloys. The presence of the metastable phase eliminates supersaturation because of its high rate of growth, and thus the growth of the stable phase is inhibited.

The electron micrographs (Fig. 73) show that the graphite layer consists of crystallites with very different shapes. As a rule, their size increases with decreasing cooling rates, although the shape of the graphite layers does not change in any significant way in the range of cooling rates investigated.

In many cases not all the grains of the metal matrix are covered with thick graphite crystals. Examination under the microscope gives the impression that some grains (whose number increases with decreasing concentrations of carbon in the alloy) are not covered with graphite at all.

However, with the electron microscope one can see that these grains as well as separate areas of grains between large graphite crystals are covered with very thin films (Fig. 73) consisting of dispersed graphite precipitated during the cooling of the alloy (apparently a carbon black type of carbon).

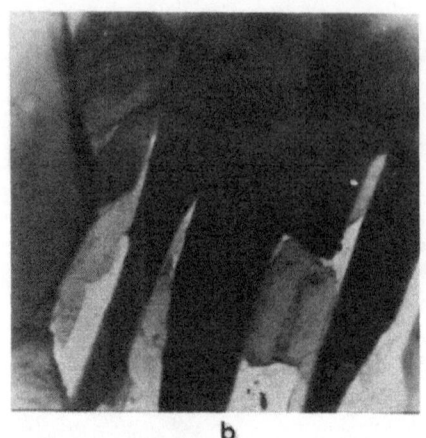

a   b

Fig. 73. Graphite due to precipitation. × 5000.

Electron-diffraction studies of graphite layers obtained as the result of cooling the alloy from the single phase equilibrium temperature indicate that the basal planes (002) of the graphite crystallites are parallel to the surface of the sample. This is indicated by the lack of reflections corresponding to the basal planes (Fig. 74a). The electron-diffraction diagram shown in Fig. 74b indicates that large graphite crystallites have some azimuthal orientation.

These results on the orientation of graphite agree with the results of x-ray studies.

In what follows we describe our investigation of the formation of graphite layers during tempering of quenched samples of Fe–C, Co–C, and Ni–C alloys. It was found, first of all, that quenching in mercury prevents the formation of graphite layers visible under the electron microscope. Nor do surface layers of graphite which can be detached electrolytically or mechanically form on quenched Fe–C alloys during tempering between 400 and 700°C.

Graphite layers visible with an optical microscope form on quenched and polished (but not etched) samples of Ni–C (0.45% C) and Co–C (0.55% C) alloys as the result of tempering at 350-650°C for 10 min and cooling at ~10°/min. Under the microscope these films appear to be uniform light-yellow or light-gray. Graphite layers also form at lower tempering temperatures, but during subsequent handling they curl or break into little pieces. Identical layers were later found to exist on the surface of slowly cooled pre-eutectoid steel (0.3% C).

The structure of graphite layers was revealed by electron-diffraction studies. Graphite layers formed at low tempering temperatures (350-400°C) consist of very small crystallites with no specific orientation with respect to the surface of the sample, as can be seen in Fig. 75a. With increasing tempering temperatures a large number of graphite crystallites with the basal planes parallel to the plane of the sample appear on the surface. This is shown by the fact that the intensity of the (002) line decreases considerably compared to the intensity of the (100) line (Fig. 75b). This electron-diffraction diagram also indicates that with increasing tempering temperatures the graphite layers on Ni–C alloys acquire an azimuthal orientation, apparently due to the appearance of large crystallites. Apparently the thickening of the layers, which can be seen under the electron microscope, is due to the graphite crystallites in which the basal planes are parallel to the surface of the sample.

The graphite layers on Co–C alloys tempered in the same temperature range do not always show azimuthal orientation. However, the number of graphite crystallites with the basal planes parallel to the surface of the sample increases with the tempering temperature.

We also investigated the structure of a graphite layer occurring during tempering of etched Ni–C and Co–C alloys. Electron-diffraction diagrams of these graphite layers indicate that etching leads to the occur-

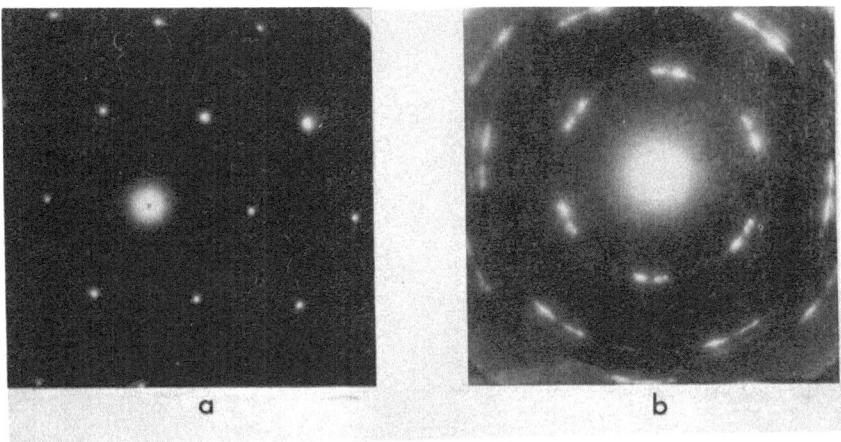

Fig. 74. Electron-diffraction diagram of graphite precipitated on austenite.

rence of texture in the (002) line corresponding to the regular orientation of graphite crystallites in which the basal plane is not parallel to the surface of the sample. Apparently etching denudes the blocks on which graphite centers nucleate; these blocks are connected to the matrix at a given angle with respect to the surface of the sample.

The analysis of the variation of the intensity of the interference rings on the electron-diffraction diagrams (Fig. 75a) indicates that low-temperature tempering of Ni–C and Co–C alloys (about 350°C) leads to precipitation of carbon black. These electron-diffraction diagrams show the presence of lines corresponding to two-dimensional diffraction.

This can be seen also in the fact that the interference maxima (hko) are asymmetrical, while the (ool) maxima are symmetrical. With increasing tempering temperatures (~500°C), interference rings corresponding to three-dimensional diffraction (112) occur on the diagram (Fig. 75b). Further increase of the temperature leads to the formation of large oriented crystals whose electron-diffraction diagrams are the same as those shown in Fig. 74.

These results obtained by electron diffraction and electron microscopic investigations of graphite precipitated on the surfaces of samples of supersaturated solid solutions of Fe–C, Co–C, and Ni–C alloys confirm the x-ray investigation, although the electron microscopic investigation gives a much clearer and more detailed picture of phase transformation.

On the basis of these results we can describe the mechanism of the precipitation of graphite on the surfaces of alloys as follows. The solubility of carbon in the solid solution decreases during cooling, and as the result graphitization centers form and grow on the surface. Under these conditions the degree of supersaturation is relatively low and the number of centers is small. Quite often, whole crystal faces of the mother phase are covered with films consisting of a few graphite crystallites. Large graphite crystallites usually are oriented with their basal planes parallel to the surface of the sample.

With increasing degrees of supersaturation (during tempering of quenched alloys) the number of graphitization centers increases and their orientation with respect to the substrata changes. At this stage the free energy of the supersaturated solid solution is so high that the formation of somewhat irregularly oriented graphite crystals becomes possible.

The thickness of the graphite film on different faces of the mother phase is uneven, since the graphitization centers are formed earlier on faces which are more favorable for their nucleation, and therefore they have time to grow to greater size during the same cooling time.

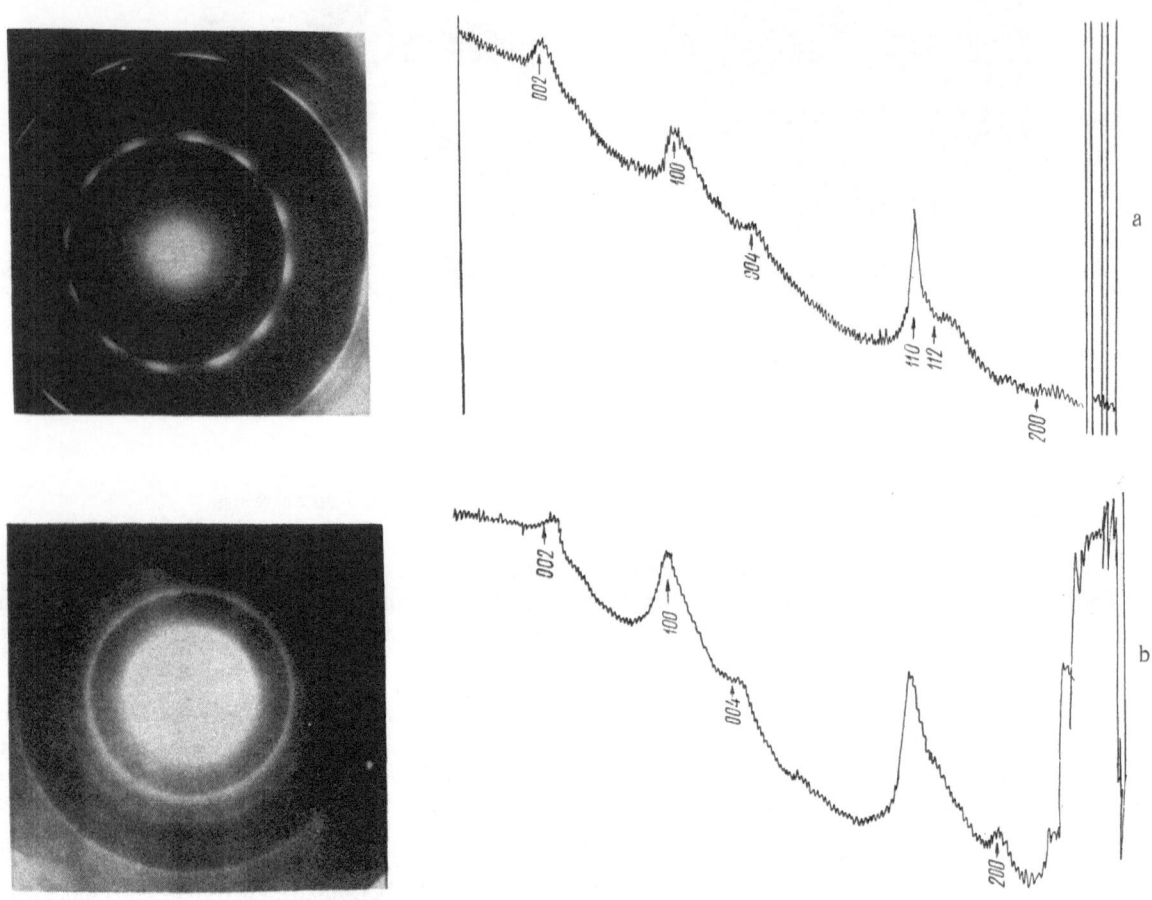

Fig. 75. Electron-diffraction diagrams and the variation of the rate of precipitation of graphite during tempering of Co—C and Ni—C alloys. a) Tempered
at high temperature; b) tempered at low temperature.

The study of the effect of the degree of evacuation on the precipitation of graphite on the surface showed that when the vacuum is $5 \cdot 10^{-1}$ mm Hg the precipitation of graphite on steel containing 1.5 % C stops. The surface of the sample becomes matte, and sometimes a blue oxide film is formed. It is clear that the precipitation of graphite on the surface is not related to the pressure as such, since graphite can easily precipitate at atmospheric pressure in a medium of an inert gas (helium, argon). The principal factor is apparently oxygen. With increasing amounts of oxygen an oxide layer with low surface tension at the boundary with the medium can form at the surface of austenite during heating, and consequently, from the energy standpoint, the precipitation of graphite is not advantageous. A second effect of oxygen is that the carbon precipitates in the atomic state during cooling of austenite, and consequently combines with oxygen, forming CO. This assumption was confirmed by the investigation of the precipitation of graphite on the surface of posteutectoid steel of the same composition which was heat treated in hydrogen and nitrogen atmospheres.

It turns out that in these media (where oxide films are not formed) graphite does not precipitate on the surface. Apparently, the carbon which precipitates in the atomic state forms compounds with these gases.

It should be noted, however, that the effect of these gases is significant only when their concentrations are high, i.e., the pressure is several cm Hg, while oxygen inhibits the precipitation of graphite even at a pressure of 0.1 mm Hg.

118

Also, oxygen reacts with iron atoms even during heating, forming stable oxide films. Therefore, oxygen is the greatest barrier to the precipitation of graphite on a ready made surface.

It is well known that graphitization does not occur in unreduced cast iron and steel because oxide films form on the inner surfaces of porosities and microcracks, which prevents the precipitation of graphite during cooling.

Let us now investigate the effect of alloyed elements on the precipitation of graphite on the surface.

It is well known that the addition of alloyed elements affects the rate of growth of precipitating phases. Therefore, we investigated the effect of V, Cr, Mo, W, Ni, Co, and Si on the precipitation of graphite from austenite during heat treatment in vacuum.

Samples containing different amounts of one of these alloyed elements and about 1.5% C were polished and heated in the vacuum furnace to 1100°C. They were kept at this temperature for 1 h and then cooled at the rate of about 30°/min. After this heat treatment they were examined under the microscope and subjected to x-ray or electron diffraction analysis. The amount of graphite precipitated was determined visually by the thickness and by the area covered.

Numerous observations of the graphitization during the heat treatment in vacuum showed that the graphite precipitating on the surface is a good indicator of the capacity of alloys for graphitization, which in itself is another advantage of vacuum metallography.

Effect of Vanadium and Chromium. We studied the precipitation of graphite on the surfaces of samples containing 1.5% C and 1, 2, and 4% V. The graphite precipitates only on samples containing less than 1% V; microscopic examination revealed no graphite on the surfaces of samples containing 2% V. X-ray and electron-diffraction studies confirmed the absence of graphite. The increase of the amount of vanadium to 4% has no effect on the precipitation process.

Thus, vanadium inhibits the precipitation of graphite and, when the amount of vanadium is more than 1%, stops it completely. Chromium also inhibits precipitation of graphite on the surface and there is no precipitation when the amount of chromium is about 2%.

Effect of Molybdenum. Molybdenum in amounts of 1-4% in alloys containing 1.5% C continuously decreases the amount of graphite precipitated on the surface, and no graphite precipitates when the amount of molybdenum is greater than 4%. The shape and character of the graphite precipitated on the surface of these samples vary in the same way as those on samples containing 1% V. Thus, an increase in the amount of molybdenum also inhibits surface graphitization, although molybdenum has less effect than vanadium and chromium.

Effect of Tungsten. The addition of 1-8% W to alloys containing 1.5% C has no effect on the amount of graphite precipitated on the surface, but the size of graphite crystals increases with increasing amounts of tungsten.

Effect of Cobalt, Nickel, and Silicon. The addition of 1-8% of these elements to alloys containing 1.5% C increases the amount of graphite precipitated on the surface. The surfaces of the samples are almost completely covered with a thick layer of graphite consisting of large polyhedral crystallites.

If we assume that the graphite precipitated on the surface is an indication of the tendency of the alloyed austenite to graphitization then the elements affect graphitization in the following order: Si, Ni, Co, W, Mo, V, Cr.

The elements to the left of the bar increases graphitization; the elements to the right decrease graphitization.

Silicon, nickel, and cobalt do not form carbides, and do not enter into the composition of graphite inclusions. Therefore, during the precipitation of graphite on the surface they do not prevent the growth of graphite nuclei. However, the growth of cementite nuclei is inhibited because atoms of these elements accumulate at the crystallization front, and therefore the growth rate of cementite inclusions depends to a great extent on the rate of removal of these elements from the crystallization front.

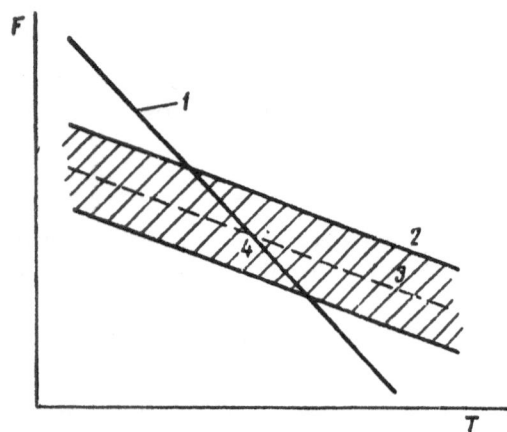

Fig. 76. Effect of alloyed elements on the free energy of metastable systems. 1) Austenite; 2) austenite + cementite; 3) austenite + graphite; 4) austenite + special carbide.

Thus, these elements have no significant effect on the mutual positions of the E'S' and ES lines, but favor graphitization because the growth rate of graphite nuclei becomes higher than the growth rate of cementite nuclei. These elements decrease the solubility of carbon in austenite, and therefore can have no great effect on the mutual positions of the E'S' and ES lines (since they are almost insoluble in cementite and graphite) and thus do not change the compositions of these phases.

The effect of the elements of the second group is somewhat more complex. These elements enter into the composition of cementite, forming alloyed cementite, and at high concentrations form special carbides. They are almost insoluble in graphite, and therefore they affect the mutual positions of the E'S' and ES lines. It is quite clear that an increase in the concentration of one of these elements in alloyed cementite increases the stability of the carbides, and consequently decreases the solubility of the carbide in austenite.

The greater the mobility of the atoms of these alloyed elements in the matrix, the greater the amount of alloyed elements in the cementite nuclei. However, the stability of alloyed cementite is greater than that of unalloyed cementite not only because the amount of alloyed elements is greater but because the stability is due to a great extent to the affinity of the alloyed elements for carbon. The greater the affinity of the alloyed elements for carbon as compared to iron, the greater the stability of the alloyed cementite as compared to that of Fe_3C.

Thus, the alloyed elements of the second group affect the relative positions of the dashed and solid lines on the Fe–C phase diagram because they change the physicochemical properties of the cementite alone. However, these elements have no significant effect on the ratio of the growth rates of cementite and graphite crystallites because their atoms are captured by the growing cementite nuclei.

The mobility of these elements in austenite * increases in the following order [191]: W, Cr, V, Mo. Their affinity for carbon increases in the following order: Cr, W, Mo, V.

Tungsten atoms increase considerably the stability of alloyed cementite, but because of their low mobility the number present in cementite is small. Apparently, only a large number of tungsten atoms can stop the precipitation of graphite on the surface. Molybdenum increases the stability of alloyed cementite and prevents graphitization because of its rather high mobility. Chromium sharply inhibits graphitization because of the high mobility of its atoms and its relatively high affinity for carbon.

The affinity of vanadium for carbon is the highest of all these elements, but at the same time it has a high mobility, which makes it an active inhibitor of graphitization.

When the amount of chromium or vanadium is more than 2% the mutual positions of the E'S' and ES lines is changed to such an extent that the ES line is situated above the E'S' line.

From the thermodynamic viewpoint these results can be represented on a free energy-temperature diagram. The lower line on the diagram (Fig. 76) indicates the variation of free energy of the austenite + special carbides system with temperature, while the dashed region indicates the possible positions of the austenite + alloyed cementite line.

Thus, the effect of alloyed elements on the precipitation of graphite from austenite is due essentially to the change in the mutual positions of the E'S' and ES equilibrium lines on the Fe–C phase diagram and also to the change in the ratio of the growth rates of graphite and cementite nuclei on the surface.

---

* Data on diffusion of vanadium are only approximations.

120

Fig. 77.   Graphite on the surface of the ingot. ×300.

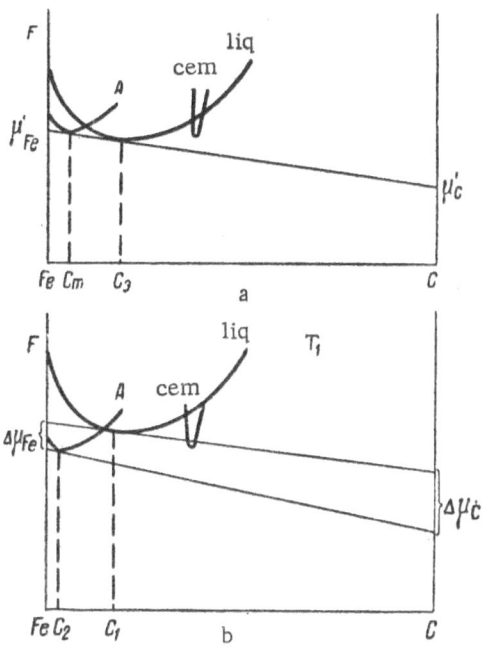

Fig. 78. Hypothetical F(TC) curves for the Fe−C system. a) Eutectic temperature of the austenite-graphite system; b) $T_1 < T$.

These experimental results agree essentially with the known effects of alloyed elements on the graphitization of cast iron and steel and they indicate that ready made surfaces (microporosities, microcracks, and other defects) play a considerable role in the graphitization process.

The graphite precipitated on the surface of austenite and ferrite when they are supersaturated with carbon can be called, by analogy with cementite formation, secondary and tertiary carbon of precipitation. If we continue the analogy, then primary graphite of precipitation must exist. This primary graphite is formed on the surface of the liquid during its solidification. This graphite actually does occur on ingots of alloys capable of graphitization (Fe−C, Ni−C, Co−C).   To prevent the masking of the precipitation of primary graphite by the formation of secondary graphite during cooling of the ingot, we cooled the liquid at a very high rate. The cooling rate was so high that the inner portions of the Fe−C ingot consisted of white cast iron.

Graphite formed on the surface of cast iron poured into a flat copper mold 5 mm deep in vacuum (Fig. 77). The amount of graphite formed on the surface depends on the composition of the cast iron and on the cooling rate.

It was shown that graphite in the solid state appears before austenite in pre-eutectoid cast iron and before cementite in posteutectoid cast iron. It forms on the surface of liquid cast iron. The formation of graphite crystals on the surface of liquid cast iron of pre-eutectic composition is followed by the crystallization of a thin layer of austenite and then a thin layer of ledeburite; only after this does normal nonequilibrium crystallization

121

continue. Crystallization of white cast iron follows the ordinary mechanism at a certain distance from the graphite film. The thickness of the austenite layer is of the order of 0.01 mm.

Silicon increases the amount of graphite formed on the surface, while chromium sharply decreases it. Thus, when the amount of chromium is as low as 1.5% no graphite whatever precipitates, and white cast iron crystallizes by the ordinary mechanism. The amount of precipitated graphite increases with increasing amounts of carbon.

In ingots of Co–C and Ni–C alloys the same phenomenon is observed when the amount of carbon exceeds the limit solubility at the eutectic temperature (C_m).

An increase in the cooling rate leads to a decrease in the amount of graphite precipitated on the surface. The precipitation of graphite on the surface of eutectic cast iron (without Si) is the easiest to inhibit. No graphite precipitates in ingots 1 mm thick. In Co–C alloys the precipitation of graphite is eliminated only when films are solidified at super-high rates (described in Chapter 6). In Ni–C alloys graphite inclusions appear on the surface of films of nearly eutectic composition even when the cooling rate is about 10^5 deg/sec.

The precipitation of graphite on the surface of solidifying liquids of posteutectoid alloys can easily be explained by the double phase diagrams, using the considerations given in section 2 of this chapter. This is ordinary primary graphite formed on the surface of the ingot because of the particularly favorable conditions for the formation and growth of nuclei. Therefore, it can be called primary carbon of precipitation.

What is the reason for the precipitation of graphite on the surface of liquids of pre-eutectic composition, which is formed earlier than the primary solid solution?

We shall use the hypothetical free energy curves to explain this phenomenon. Figure 78 shows the position of the F(CT) curves for Fe–C alloys at temperatures T and T_1. The temperature T_1 is below the eutectic temperature. At this temperature austenite with a composition $C_2 < C_m$ (C_m is the limit stable solubility of carbon in austenite) can precipitate from the liquid with a composition $C_1 < C_{eu}$. It is easy to see that the graphite nuclei and not the austenite nuclei will be formed first because $\Delta\mu_{Fe} < \Delta\mu_C$.

In accordance with our previous discussion, the most favorable place for the formation of graphite nuclei is the surface. Austenite is formed in the liquid after the graphite is formed. The simultaneous growth of the graphite and the austenite continues until the concentration of the liquid at the crystallization front of the austenite layer becomes such that cementite nuclei can form in it. Then, ledeburite colonies begin to grow at a very high rate and thus stop the growth of the surface film of primary carbon of precipitation.

A higher degree of supercooling leads to simultaneous formation of graphite and cementite nuclei. Since the growth rate of cementite nuclei is greater than that of graphite nuclei, they inhibit the growth of graphite nuclei and completely stop it at the stage where graphite inclusions are not visible even under relatively high magnification. Much higher degrees of supercooling are necessary to prevent the precipitation of primary graphite in Co–C and Ni–C alloys.

Thus, the characteristic singularity of alloys capable of graphitization is that when the cooling rate is high they form a metastable carbide on solidifying. However, the singularities of the structure of the stable phase are such that it is very difficult to inhibit the formation of nuclei on different types of ready made surfaces by increasing the cooling rate alone. Therefore, unalloyed alloys of this system contain a large number of relatively large graphite nuclei when the cooling rate is high.

CHAPTER 9

TRANSFORMATION OF THE METASTABLE SYSTEM
INTO THE STABLE STATE

1. The Stability of Metastable Systems

Different meanings have been given to the term metastable state. On the basis of a strict thermodynamic definition of metastability, almost all systems are in the metastable state most of the time. In most cases it is for practical purposes impossible to reach an absolute minimum energy.

A liquid is transformed into a solid only after the original liquid is supercooled, i.e., is transformed into the metastable state. When the degree of supercooling is low, the liquid can remain in the metastable state indefinitely. Only when the degree of metastability is increased (when the degree of supercooling is increased) is the system finally transformed into a system with a lower free energy. However, even here there is a hidden contradiction. The higher the degree of metastability of the initial state, the higher the degree of metastability after the transformation. In fact, the higher the degree of supercooling, the greater the dispersity and the branching of the crystals of the new phase. Sometimes this transformation is accompanied by internal stresses and the formation of vacancies. The simplest method of reaching a state with minimum free energy is the use of seed crystals and very low cooling rates.

When the liquid is supercooled in the presence of a ready made crystal of a new phase and the cooling rate is extremely low, then a single-crystal system is formed and the stacking of atoms at the separation boundary is the most perfect. In other words, the whole system has minimum volume and surface energies. A system obtained in this way must be considered stable, since it possesses a minimum free energy, and all the other stable states of the system must be considered metastable.

In some alloys there may exist several intermediate metastable states each of which has a characteristic value of free energy. The order in which these intermediate states are formed during supercooling of the initial phase is sometimes believed to follow the so-called "Ostwald step rule." According to this rule, the initial phase is first transformed into a metastable phase and later (at a lower temperature) into a more stable phase. It was noted in [259, 260] that this is not really a rule. We have shown previously that the amount of the metastable phase formed is determined in each particular case by the ratio between the formation rate and the growth rate of nuclei of a given phase. In many cases, stable and metastable phases are formed simultaneously, in contradiction of Gibbs' rule.

Thus, metastable system means a number of intermediate states differing from each other not only in the values of the free energy but also in the possibility of transformation into the stable state.

The mechanism of the transformation into the stable state is very important, since it determines the degree of stability of a given metastable system. Metastable systems occur only when supercooling results in the formation of a system with a high free energy, since when a unicomponent liquid is supercooled above the metastability limit it is in a supercooled state because the activation energy necessary for the formation of the crystal-liquid boundary is still too high. Further decrease in temperature leads to the formation of nuclei, and consequently to the overcoming of this first energy barrier.

In some cases the nuclei grow as the result of the formation of two-dimensional nuclei and in other cases as the result of the combination of single atoms. In the first case, the system is overcoming an energy barrier equal to the work of formation of a two-dimensional nucleus, and in the second case the energy barrier is equal to the activation energy of the motion of atoms in the liquid.

The formation of metastable systems as the result of the decomposition of supersaturated solutions is characterized by the same types of energy barriers as in the case of transformation into the stable state.

The excess free energy is the measure of metastability because it characterizes the deviation of the system from a state with minimum free energy, but it cannot be a criterion of the stability of a given system, since the stability of a system is determined by kinetic factors.

The free energy of the activation of the transformation process is such a universal criterion. In this case it is the minimum energy necessary for the elementary acts of the transformation of this system from the metastable to the stable state.

Depending on the mechanism of the transformation into the stable state, metastability can be divided into the following types:

1. Metastability due to the work of formation of three-dimensional nuclei of a new (stable) phase. All the phase transformations in the solid, liquid, and gaseous states in which crystallization begins with the formation of nuclei belong to this type of metastability. Transformations in the solid state have some singularities, but these are taken into account in determining the work of formation of nuclei. Usually, the activation energy at the beginning of such transformations decreases with decreasing temperature. At the same time, the decrease of the temperature leads to an increase in the metastability of the final system resulting from the transformation.

2. Metastability due to the work of formation of two-dimensional nuclei. This is a very rare case in alloys. It occurs during the growth of single crystals when the growth of perfect faces of the seed crystal is inhibited because the supercooling is insufficient for the formation of two-dimensional nuclei. Single crystals grow under conditions which exclude the formation of three-dimensional nuclei.

3. Metastability due to the activation energy of the displacement of single atoms. In many cases this energy is close to the diffusion or self-diffusion activation energy of atoms. When the lattice of the system which has undergone polymorphic transformation undergoes coherent rebuilding, the activation energy can be even lower, since the distances the atoms move are less than the interatomic distances. Recrystallization processes belong to this type of metastability.

4. Metastability due to the energy of the motion of dislocations. This type of metastability exists in elastically stressed systems. Usually, the relaxation of stresses is related to the motion of dislocations, which requires a very low activation energy in some cases. Problems to do with the relief of stresses are usually dealt with in the theory of relaxation phenomena.

By changing the cooling rate one obtains one or the other type of metastability and, depending on the value of the activation energy, an alloy in a given state will have a given degree of stability.

Most of the solid metastable systems existing at low temperatures are transformed into the stable state during annealing in a rather complex way.

In a supersaturated solid solution the formation and growth of nuclei of the second phase are accompanied by the creation and relaxation of stresses, coalescence, change in the shapes of growing centers, etc. Each of these phenomena affects in some way the kinetics of precipitation and the properties of the alloy.

The transformation of a system containing intermediate metastable phases into the stable state is even more complex. Among the systems with intermediate metastable phases are cast iron and alloyed steels forming cementite and special carbides.

The stability of Fe−C alloys in which the high-carbon component is iron carbide is determined by the graphitization rate, i.e., by the time it takes for the carbide to dissolve and the graphite to form. Up to now there has been no agreement on the so-called narrow region in which the graphitization process occurs.

The narrow region in which graphitization begins and which essentially determines the activation energy of this process is related to the nucleation of graphitization centers and, consequently, is determined by the work of formation of graphite nuclei. Further growth of graphite nuclei can be inhibited by the diffusion of carbon, the dissolution of carbides, etc. If we add to all this the effect of the alloyed elements, then the complexity of the mechanism by which the system is transformed into the stable state is increased many times.

The decomposition of martensite and austenite in quenched simple and alloyed steels is as complex. Several types of carbide are formed, some of them stable and some metastable.

Unfortunately, most investigations of these transformations have been made with complex systems in which many factors interact simultaneously, and thus mask the general relationships characteristic of transformations into the stable state in a wide range of systems.

To clarify the determining factors in the initial stages of transformation into the stable state it is first necessary to know whether the centers of the stable phase are formed during the process of transformation or during the formation of the metastable state (and, consequently, whether the transformation results from the growth of ready made nuclei). Experimentally, this is a very complicated problem. However, no doubt the development of electron microscopy will lead to solution of this problem in the near future.

It is also necessary to determine the degree of stability of each metastable phase. During subsequent heating or annealing, can a metastable phase decompose spontaneously into its components or into other stable phases ?

From the thermodynamic viewpoint, Fe−C alloys containing cementite are metastable in the entire range of concentrations from Fe to Fe_3C and from Fe_3C to C (see Fig. 78b). However, in many cases they are extraordinarily stable. Recently, the spontaneous decomposition of isolated cementite bombarded with elementary particles in vacuum was investigated [262, 263]. It was shown that cementite decomposes, although very slowly.

On the basis of the results obtained it is still difficult to decide whether the decomposition of cementite in steels and cast irons is spontaneous, since isolated cementite is a completely independent thermodynamic system differing from the Fe−C system.

Information is also needed on the stability of special carbides coexisting in austenite and graphite.

The time necessary for the transformation of metastable systems into the stable state is taken up mostly by diffusion processes. In systems containing intermediate metastable phases the transformation time is particularly long; the time depends simultaneously on the dissolution of crystallites of the metastable phase and the growth of centers of the stable phase. Sometimes it is assumed that the stability of such systems is determined by the dissolution energy of the metastable phase in the solid solution. It is the difference in the solubilities which creates the necessary concentration gradient as a result of which the substance diffuses from one phase into another phase. At this stage the mobility of the atoms of the components is also very important.

The general laws of the transformation of such complex systems as alloys with intermediate phases into the stable state cannot be described at present. Some important singularities of this phenomenon will be described below in the case of cast irons and steels capable of graphitization.

In the case of single-phase decomposition of a supersaturated solution the processes sometimes stops in the early stages. The process of decomposition stops when the precipitated phase is still highly dispersed, when the solid solution is highly supersaturated with respect to the equilibrium concentration. The stability of a metastable system under these conditions was investigated in detail in [204]. It was found that there is colloidal equilibrium in this case. The stability of the colloidal system is determined by the fact that the dispersed particles are almost the same size. Under these conditions $R_0 - r_0 \approx 0$ and the growth constant a also becomes zero in Eq. (7.7), and consequently coalescence ceases. Dispersed particles cannot grow because of the decomposition of the solid solution, since the amount of dispersed phase precipitated decreases the concentration to the equilibrium concentration, which is determined by the Thompson equation.

2. Rate of Transformation into the Equilibrium State

Different forms of metastability may arise in a crystallizing system, depending on the degree of supercooling. Among these forms of metastability are the fragmentation and branching of crystallites of different phases, formation of supersaturated solutions, and finally the creation of metastable phases. In all these cases the transformation into the stable state is related to changes in the chemical potentials of atoms of the components and a necessary decrease in the free energy of the system.

Any change in the composition of the phases or in the shape of the crystals is related to displacement of atoms of the components. Therefore, the rate of these transformations is determined by the rate of motion of

125

atoms of a given type in the direction of the acting force $\overline{f} = b(\partial\mu/\partial x)$. Here, the coefficient b is the mobility of atoms during self-diffusion or the rate of thermal motion of atoms when the concentration gradient is zero at a given temperature. According to Einstein, the value of b is expressed as b = D*/kT, where D* is the self-diffusion coefficient of atoms of a given type in an equilibrium solution. The direction of the motion of atoms is determined by the sign of the operator $\nabla\mu$. Thus, other conditions being equal, the rate of motion of atoms depends on the gradient of the chemical potentials of atoms of the components in a given area of the alloy. In an ideal case the rate of motion of atoms is proportional to the concentration gradient. Since the activity coefficient f = 1, then

$$\frac{\partial\mu}{\partial x} = \frac{kT}{C}\frac{\partial C}{\partial x}.$$

In real alloys we have

$$\frac{\partial\mu}{\partial x} = \frac{d\ln a}{d\ln f}\cdot\frac{kT}{C}\frac{\partial C}{\partial x},$$

and the diffusion coefficient of atoms of any component is determined by D = D*(d ln a_i/d ln f_i), and consequently depends on the activity coefficient. This relationship is very important in cases where transformations are accompanied by considerable changes in the composition and density of the system. In the latter case vacancies are formed; they play the role of a third component and have considerable effect on the activity coefficient. The term following D* is called the thermodynamic multiplier. If the binding energy between the atoms of the components is greater in the solution than between the atoms of the pure components then the thermodynamic multiplier is greater than unity, and consequently the atoms of a given component will diffuse more rapidly than in a homogeneous solution. If the thermodynamic multiplier is less than unity their diffusion will be slower than in a homogeneous solution.

All this refers to the rate of motion of atoms in the crystal lattice of the solid solution at a given temperature. However, atoms pass from one equilibrium position to another by jumps, and therefore the diffusion coefficient itself depends also on the value of the activation energy Q, which to some degree makes it possible to judge the relative force of atomic interaction in different lattices.

All this indicates that the rate of transformation of a system into the equilibrium state at a given concentration gradient is determined by a complex (effective) diffusion coefficient. The value of this diffusion coefficient may be very different from the values given in the tables. This is particularly true in the case of diffusion of atoms of the components of substitution solid solutions. Thus, if one solid solution is transformed into another, then the partial diffusion coefficients of components A and B in the case of the vacancy mechanism are determined in the following way:

$$D_A = D_A^*\left(1 + \frac{d\ln a_A}{d\ln f_i}\right);\ D_B = D_B^*\left(1 + \frac{d\ln a_B}{d\ln f_B}\right),$$

and the general diffusion coefficient [257, 258] is determined by

$$D = D_A(1 - C_A) + D_B C_A.$$

In the case of formation of carbides in steels or cast irons one must take into account not only the diffusion of carbon but also the diffusion of iron, and in the case of alloyed steels and cast irons one must also take into account the diffusion of the alloyed elements.

When several diffusional flows interact the reaction rate is determined not by the slowest of them but rather is a complex function of all these different diffusion coefficients.

In all these cases the surface tension plays a very important role. In many cases the density of the alloy changes when the system is transformed into an equilibrium state and this change of density is accompanied by plastic deformation, formation of vacancies, dislocations, porosities, microcracks, etc. All this changes sharply the value of the surface tension, which as we have seen determines the kinetics of the process to a considerable extent.

Diffusional growth of the phase as well as its nucleation is connected to the passage of atoms from one energy state into another. Therefore, in both cases one can use the activation energy Q as the value which in some approximation is dependent on the temperature. Then, by analogy with the diffusion process, the rate of transformation into the stable state as a function of temperature can be represented as

$$v = v_0 e^{-\frac{Q}{RT}}$$

or

$$\ln v - \ln v_0 - \frac{Q}{RT}.$$

This relationship is often used in comparing the kinetics of graphitization of cast iron and steel under different annealing conditions, the effect of alloyed elements, the kinetics of the decomposition of martensite, and even to determine the activation energy of the martensitic transformation itself.

The value of Q is determined by the slope of the line representing the variation of the reaction rate with the temperature. In some cases the activation energy determined in this way is ascribed to the passage of atoms through the phase separation boundary, and in other cases to the diffusional displacement of atoms in the lattice of the solid solution.

The activation energy of the transformation into the stable state is a complex characteristic which depend. on many factors [217].

The activation energy of diffusionless transformation of austenite into martensite in carbon steel is much lower than the activation energy necessary for the formation of cementite nuclei or carbon diffusion in the lattice of iron. Nevertheless, martensite cannot be reversibly transformed into austenite, regardless of the heating rate, since martensite has time to decompose partially [261], apparently because austenite is transformed into martensite at low temperature, when lattice reconstruction is coherent. In these cases the rate of transformation is extremely high.

The reverse transformation of martensite into austenite occurs at a much higher temperature, when the thermal oscillations inhibit the orderly passage of atoms from martensite into austenite. The transformation rate decreases sharply and the rate of decomposition becomes commensurable with the rate of rebuilding of the lattice of the α-solution into the lattice of the γ-solution.

Thus, the rate of transformation of a complex metastable system into the stable state is determined by the kinetics of the process which occurs at each given stage of the transformation.

Among these processes are nucleation and growth of centers, coalescence, stress relaxation, spheroidization, and polyhedrization.

3. Carbon Due to Annealing

Bunin and Salli [256] showed for the first time that the graphitization of white cast iron begins earlier on the surface than within the samples. Further study of the graphitization process in white cast iron revealed the principal factors in the mechanism and the kinetics of this phenomenon.

Most of the investigation concerned pre-eutectic and posteutectic cast iron.

The compositions of the cast iron were : 1) 4% C, 0.1% Mn, 0.5% Si; 2) 3.5% C, 0.3% Mn, 1.2% Si; 3) 3% C, 0.1% Mn, 0.5% Si; 4) 4% C, 0.1% Mn, 0.5% Si, 2% Cr; 5) 4% C, 0.1% Mn, 0.5% Si, 3% Cr.

The graphitization process on the surface of white cast iron was investigated in the following way.

The surfaces of the white cast iron samples were polished, placed in the vacuum furnace, and heated. The samples were then cooled to room temperature in the furnace and examined under the microscope.* The graphitization of white cast iron was observed during solidification of ingots smelted in a vacuum furnace, using the method described in the preceding chapter.

---

* The surfaces examined were cross sections of the original ingot. Most of the ingots were flat or cylindrical. Thus, in what follows we are speaking of the cross section unless we specify that it is the surface (crust) of the original ingot.

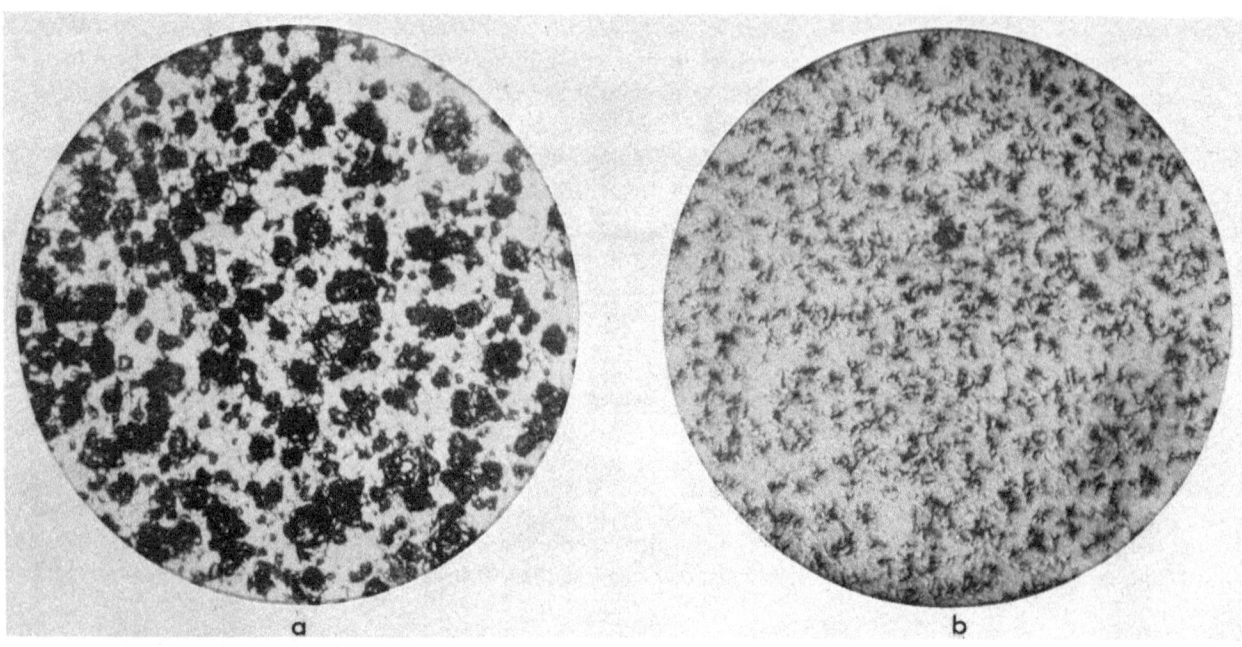

Fig. 79. Carbon due to annealing on the surface of white cast iron. a) Cast iron containing 4% C and 0.5% Si after 25 min at 950° C;   b) same cast iron after polishing with a cloth.

Graphitization centers formed on the polished surface of white cast iron heated in vacuum (Fig. 79a).

In the beginning these centers are hardly visible and their number is very small. Then they grow in numbers and at the end are so large that they completely cover the surface of the sample.

If the surfaces of such samples are lightly polished one finds ordinary nests of carbon due to annealing beneath the graphite layer (Fig. 79b). A thorough investigation of the places where the graphitization centers occur showed that these centers never occur on cementite grains but always occur at the point where the cementite grains join or at boundaries with austenite. It is quite interesting that these centers always occur only in the ledeburite of eutectic or pre-eutectic cast iron.

Unlike the surface graphite previously described, this graphite (ordinary carbon due to annealing) grows in the form of gray protrusions often composed of large crystals. The nests occupied by the graphite inclusions in the metal matrix are shallow and often branched depressions which are smaller than the graphite inclusions themselves. The nests of carbon due to annealing and the graphite on the surface are shown in cross section in Fig. 80. This structure of inclusions of carbon due to annealing formed on the surface is easily explained by the fact that the specific volume of graphite is very large, while the nest formed as the result of the diffusion of vacancies (formed during the dissolution of cementite) to the crystallization front of the graphite center is too small to contain all the graphite. Therefore, during the formation of graphite centers the principal portion of the graphite creeps to the surface and finally covers it completely.

On the basis of the fact that the graphite covers the whole surface of the sample during graphitization of white cast iron, a completely erroneous conclusion was drawn in [255], where it was said that the graphitization centers also occur on the surface of cementite grains. The shapes of the depressions in the metal matrix under the graphite protrusion can differ greatly.

These surface processes depend on the composition and the structure of the initial alloy. The rate of surface graphitization of white cast iron is usually higher than the rate of graphitization within it. In synthesized cast iron containing a small amount of silicon, graphitization occurs essentially on the surface. It terminates after all the cementite adjacent to the surface of the sample is dissolved.

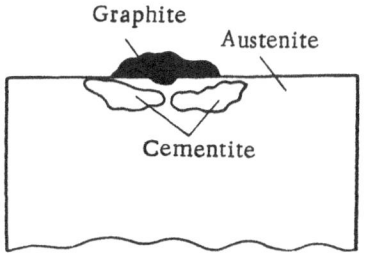

Fig. 80. Diagram of surface
graphitization.

In industrial cast irons containing large amounts of silicon the difference in graphitization rates is much smaller. In synthesized cast iron the time necessary for total graphitization of the surface and graphitization of the inner parts of the sample differs by 10 h, while the difference is only 5 min in the case of industrial cast irons.

The distribution of graphitization centers on the surface and within the sample are similar. In [265] it was shown that the distribution of graphitization centers is determined by the microstructure formed during the solidification of liquid cast iron and, particularly, during solidification of the eutectic.

White cast iron containing no silicon or containing more than 2% Cr is not subject to graphitization either on the surface or within the sample.

Examination of the surface of cast iron solidified in vacuum showed that if the composition of white cast iron is close to the industrial composition then graphitization on the surface occurs even during cooling of the ingots. The graphitization is most intense on surfaces where the cooling rate is low. In cast irons in which there is almost no graphitization within the sample there are almost no graphite centers on the surface either. In cast irons cooled at a very high rate graphitization is slow both on the surface and within the ingot.

In cast iron either quenched or kept at 300-350°C before annealing the number of graphitization centers is increased even on the surface, although this increase is somewhat less than that within the sample.

In spite of numerous investigations, no one has succeeded in observing spontaneous decomposition of cementite.

In cast iron in which there is almost no graphitization (and, consequently, there are no graphite centers) cementite does not decompose on the surface even after 200 h at 1050°C. It was also found that cementite is stable not only at the boundary with vacuum but also when it is in direct contact with graphite.

All these experimental results can be explained only if one assumes that the graphitization of white cast iron proceeds essentially on ready made graphite crystallization centers, as discovered by Bunin [239] and Bogachev [238]. These investigators found that the graphite in white cast iron is formed on a great number of graphite inclusions occurring in the liquid cast iron and also during the solidification of the ingot. Only a few of this enormous number of graphite inclusions − those which during cooling of the ingot have time to reach sizes at which they are stable at a high temperature in the surrounding cementite in ledeburite − become the graphitization centers in white cast iron.

The most favorable conditions for the formation of such stable nuclei during solidification of ingots occur when an alloyed element − which is responsible for the increasing distance between the dashed and solid lines on Fe−C phase diagrams − is added to the alloy, or when the cooling rate is low. If these conditions are not satisfied the graphite inclusions are very small and they are "absorbed" by the surrounding cementite during annealing of cast iron. However, even under favorable conditions enormous numbers of "prenuclei" are formed. These "prenuclei" can become graphitization centers of white cast iron only under conditions which favor their growth to a supercritical size. Such conditions (aside from those mentioned in section 3) occur on the surface of the sample.

The main factor responsible for the increase of the growth rate of a graphite center is a great concentration difference, a sufficient growth rate, and the existence of space for the growth of the graphite centers. The conditions are more favorable on the surface of the sample than within the sample for the following reasons.

The difference in the concentration increases as the result of evaporation of iron atoms from the surface and as the result of the presence of sharp cementite ridges, formed during polishing, which increase the solubility of carbon in the austenite at the cementite boundary. The diffusion rate of carbon is also somewhat higher on the surface than within the sample.

However, the most favorable condition for the growth of the nuclei on the surface is the fact that iron atoms do not have to move from the crystallization front, since the nuclei grow essentially parallel to the surface. The nests in the austenite become wider because of the diffusion of vacancies from the dissolving cementite.

As the result, the graphitization rate on the surface is much higher than within the sample, and the more difficult the graphitization of white cast iron the greater the difference. In cast iron containing a large amount of silicon, graphitization develops on a great number of large graphite inclusions on the surface and within the sample, and therefore the whole process takes a short period of time.

It is interesting that about the same number of graphitization centers are formed in the crust and on its surface. The growth of these centers is somewhat higher on the surface, which indicates that the casting surface favors the growth of large centers rather than small centers.

The most favorable conditions for the growth of small centers are ensured by quenching to martensite. This can be seen in that the quenching increases the number of centers formed on the surface of the sample.

These results are convincing proof that on the surface graphitization begins on a definite number of initial centers which depends on the preliminary treatment. When the treatment of the surface is not very different from the treatment of the inner parts of the ingot, the number of centers on the surface is almost the same as within the ingot. The same factors which inhibit graphitization on the surface (presence of chromium, absence of silicon, casting crust, etc.) also inhibit graphitization within the ingot.

These experiments confirm the fact that the graphitization of white cast iron during isothermal transformation into wrought iron results not from spontaneous graphitization of centers according to the ordinary crystallization mechanism but proceeds from ready made graphitization centers. In fact, when graphitization centers occur on the surface (graphitization of austenite) their number is large and they cover the whole surface of the austenite. The amount of graphite precipitated on the surface depends only on the degree of supersaturation of austenite with respect to the E'S' line.

In graphitized white cast iron the number of graphitization centers on the surface is directly proportional to the number of graphitization centers within the sample.

In those parts of the ingot where the number of stable graphitization centers is large within the ingot the number on the surface is also large.

If the graphitization centers occur during the annealing of white cast iron then the number on the surface would be large and they would cover the surface uniformly without forming nests in the metal matrix, as in the case of graphitization of austenite.

It would be logical to assume that the number of carbon inclusions resulting from annealing of white cast iron cannot exceed the number of already existing graphitization centers which have occurred during solidification of the cast iron.

The maximum number of carbon inclusions due to annealing is formed when the ingot is quenched before annealing (i.e., when one creates a medium favorable for the growth of the existing nuclei to supercritical size — martensite, supersaturated austenite), then is heated at 350-400°C (a temperature at which martensite and austenite decompose and the graphite nuclei can grow to supercritical size), and then is annealed to transform the ingot into wrought iron.

The quenching temperature must be relatively low and the time the ingot is kept at this temperature before quenching must be short. If the temperature is high and the time the ingot is kept at this temperature is prolonged, then large numbers of small nuclei which are unstable in the surrounding ledeburite cementite are lost.

We studied a pre-eutectic cast iron composed of 2.2% C, 1.28% Si, 0.43% Mn, 0.03% S, 0.028% P, 0.034% Cr. The cast iron was poured in a copper chill mold to obtain plates 80×40×4 mm. The plates were broken into small samples which were subjected to different heat treatments.

TABLE 17. Variation of the Number of Grains with the Heating Time

Quenching temperature,°C	Heating time, min	No. of grains per cm^3	Quenching temperature,°C	Heating time, min	No. of grains per cm^3
760	20	$1.3 \cdot 10^9$	880	20	$3.14 \cdot 10^8$
760	30	$6.4 \cdot 10^6$	880	30	$0.64 \cdot 10^6$
760	40	$1.3 \cdot 10^6$	880	40	$0.35 \cdot 10^6$

The following results were obtained:

1. The greatest number of centers results from the following heat treatment. The samples are heated to 750-800°C for no more than 20 min and quenched; then kept at 400°C for 3-4 h; then annealed at 950-1000°C.

This heat treatment results in $1.3 \cdot 10^9$ carbon inclusions per cm^3 of the sample.

2. When the samples are not subjected to preliminary quenching and not heated at low temperatures the number of carbon inclusions decreases with increasing annealing temperatures.

Thus, there are $1.33 \cdot 10^6$ grains per cm^3 when the samples are annealed at 700°C and only $0.35 \cdot 10^6$ grains per cm^3 when they are annealed at 880°C.

3. The number of carbon inclusions due to annealing decreases when the time of heating to the quenching temperature is increased (Table 17).

4. The greater the number of quenching cycles, the smaller the number of graphitization centers.

Thus, the number of inclusions per cm^3 varies with the number of quenching cycles at 780°C in the following way:

	Number of inclusions per cm^3
Quenched once...........	$1.1 \cdot 10^9$
Quenched twice	$5.9 \cdot 10^8$
Quenched three times	$5.3 \cdot 10^8$

5. Other types of heat treatment used for cast iron consistently produce a smaller number of inclusions of carbon due to annealing than the number resulting from the first heat treatment. *

These experimental results confirm the graphitization mechanism described earlier.

Thus, the main singularity of the graphitization of white cast iron is that graphitization proceeds on ready made graphitization centers formed during cooling of the melt and during the additional heat treatment, which favors the growth of numerous "frozen" (unstable) centers to a stable size. The role of "biographical" defects becomes clear. These defects are located mainly between the dendritic branches, in places where there is some looseness in the structure, etc. When cast iron is cooled, the graphite precipitated from the austenite deposits on the inner surfaces of these defects and transforms them into future graphitization centers. These centers grow whether they are on the surface of within the sample, outlining the dendritic structure. The number of graphitization centers formed on "biographical" defects is smaller in those cast irons in which austenite is difficult to graphitize. Graphitization of austenite is difficult when the cooling rate is low or when the system contains an insufficient number of graphitizing impurities. In these cases graphitization is due only to the presence of graphitization centers which were formed in liquid cast iron and had time to grow to stable size. Quenching, preliminary heating at low temperatures, etc., increase the number of nuclei. When the ingot is cooled slowly and the alloy contains a sufficient amount of a graphitizing element, preliminary low-temperature heating has no significant effect; nor does it have any effect when the ingot is taken out of the mold at high temperature.

Apparently, graphitization of steel proceeds either on centers formed on "biographical" defects during cooling or on centers formed during quenching (which grow during heating to the graphitization temperature).

---

* Excluding the increase of the heating time at low temperature after quenching.

These processes are the most common in posteutectoid steels.

Thus, the transformation of a metastable system (cementite-austenite) into a stable system (graphite-austenite) is possible because of the existence of ready made centers for the crystallization of the stable phase. The probability of the formation of new graphite centers during isothermal heating of white cast iron is small because the surface tension at the graphite-austenite boundary is greater than at the cementite-austenite boundary.

The kinetics of the transformation of a metastable system into a stable system is determined by the number of graphite nuclei existing in the system and by their growth rate which, in turn, depends on many factors, such as the concentration gradient, the rate of removal of iron atoms from the graphite crystallization front, the diffusion rate of carbon atoms and atoms of alloyed elements, etc. Each of these is a factor in the growth of graphite crystals at different stages of graphitization in different alloys.

We can cite another confirmation of the graphitization mechanism of cast iron and steel described here. No graphitization of posteutectoid and pre-eutectoid steels was observed in the study of the coalescence of cementite in carbon steel. Samples of carbon steel were kept 40 h and more in a vacuum furnace, but no traces of graphitization could be detected on the surfaces of the samples. Yet, samples of posteutectoid steel containing more than 1% C graphitize rather rapidly. Inclusions of carbon due to annealing are clearly visible on steel samples containing 1.4% C after 100 h annealing at 650° C. Quenched samples of posteutectoid steel are easily graphitized.

This phenomenon can be explained neither by the dissociation of cementite nor by spontaneous nucleation of graphite centers during isothermal heating.

The curves representing the rate of nucleation of graphitization centers during graphitizing annealing published in the literature do not prove the nucleation of graphite during annealing. The apparent increase in the size of graphitization centers in cast iron and steel is the consequence of the growth of ready made graphite centers created during cooling.

Finally, let us note that the main conclusions derived in this chapter are not limited to graphitization of Fe–C alloys. They are relevant to the transformation of many metastable systems into the stable state.

The metastable phase occurs not only because the rate of formation of its nuclei is greater than the rate of formation of nuclei of the stable phase but also because the growth rate of nuclei of the metastable phase is very much higher than that of nuclei of the stable phase.

In all systems the transformation of a metastable system into the stable state requires the presence of nuclei of the stable state. If this were not so, a metastable system could exist indefinitely.

Also the size of nuclei of the stable phase surrounding crystallites of the metastable phase is a primary factor in many systems.

This statement was confirmed by an investigation of the mechanism of the decomposition of austenite in steel containing molybdenum [266]. The authors found a number of relationships similar to those found for the graphitization of Fe–C alloys. When the austenite in an alloyed steel decomposes, the degree of supersaturation being low, a stable phase is created – the special carbide $(FeMo)_{23}C_6$.

The most interesting point is that the addition of molybdenum to steel considerably decreases the rate of decomposition of austenite only when the degree of supercooling is relatively low. As the degree of supercooling increases, metastable cementite (Fe_3C) is formed.

This result indicates that the rate of formation of the nucleus of the stable phase is much lower than the rate of the formation of cementite nuclei because the nucleus of the stable phase has a high surface tension and a higher molecular weight. In this case also, the phase diagram must have two systems of lines to indicate the phase equilibrium conditions at different temperatures and concentrations.

The growth rate of the cementite nucleus, other conditions being equal, is much higher than the growth rate of the special carbide nucleus. The first is determined essentially by the diffusion rate of carbon to the crystallization front, the second by the diffusion of the alloyed element.

Therefore, when the degree of supercooling is very high, cementite nuclei (as in the case of graphitization) "outgrow" the nuclei of special carbides. However, the nuclei of special carbides which, during isothermal heating, grow to a size ensuring their stability in a cementite environment, "absorb" the metastable cementite.

The study of the coalescence of dispersed carbides in alloyed steels revealed this mechanism of the transformation of a metastable system into a stable state.

The "absorption" of cementite inclusions by special carbide crystals was confirmed in [208]. In this case, as in the case of graphitization of steel and white cast iron, the concentration of carbon in the mother phase is lower at the boundaries with the special carbides than at the boundaries with cementite. The difference in the concentrations ensuring the growth of nuclei of special carbides is determined at first by the size of the nucleus: if it is very small it is dissolved in the presence of the metastable phase.

REFERENCES

1. Ya. S. Umanskii, B. N. Finkel'shtein et al., Physics of Metals, Metallurgizdat, 1950.
2. V. Hume-Rothery and R. V. Raynor, Structure of Metals and Alloys [Russian translation], Metallurgizdat, 1959.
3. A. H. Cottrell, Structure of Metals and Alloys [Russian translation], Metallurgizdat, 1961.
4. S. A. Vekshinskii, New Methods of Metallographic Investigation, OGIZ, 1944.
5. A. Einstein and V. Smolukhovskii, Brownian Motion, ONTI, NKTP, SSSR, 1936.
6. Esenstein and Gingrich, Phys. Rev., 46: 703 (1934).
7. G. H. Vinejrd, Phys. Rev., 62: 261 (1942).
8. R. L. Wild, J. Chem. Phys., 18: 1627 (1950).
9. N. V. Mokhov, K. E. Kolisnichenko, and Ya. M. Lobkovskii, in: Papers of the Physics-Mathematics Department of Dnepropetrovsk State University, No. 5, 1956.
10. I. Z. Fisher and K. Prokhorenko, Zh. Fiz. Khim.,33: No. 8 (1959).
11. Ya. I. Frenkel', Kinetic Theory of Liquids, Izd. AN SSSR, 1945.
12. V. I. Danilov, Structure and Crystallization of Liquids, Izd. AN SSSR, 1956.
13. K. P. Bunin, Izv. Akad. Nauk SSSR, Otd. Tekhn. Nauk, 2: 305 (1946).
14. G. M. Barten'ev, Zh. Tekhn. Fiz., 17: No. 11, 1325 (1947).
15. A. S. Lashko, Zh. Fiz. Khim, 33: No. 8, 1730 (1959).
16. A. F. Skryshevskii, in: Problems of the Physics of Metals and Metal Science, No. 8, DAN UkrSSR,(1957), p. 187.
17. N. I. Bublik and I. M. Buntar', Kristallografiya, 3: No. 1 (1958).
18. A. A. Vertman, A. M. Samarin, and A. M. Yakobson, Izv. Akad. Nauk SSSR, Otd. Tekhn. Nauk, Met. i Toplivo, No. 3 (1960).
19. A. A. Vertman and A. M. Samarin, in: Structure and Properties of Liquid Metals, Izd. AN SSSR, 1960.
20. A. F. Skryshevskii, in: Structure and Properties of Liquid Metals, Izd. AN SSSR, 1960.
21. A. H. Cottrell, in: Progress in the Physics of Metals, No. 1, Metallurgizdat, 1956.
22. I. A. Oding, Theory of Creep and Resistance of Metals to Prolonged Heating, Metallurgizdat, 1956.
23. I. A. Oding, Theory of Dislocations in Metals and Its Applications, Izd. AN SSSR, 1954.
24. "Classification and Manifestations of Residual Stresses," Zavodsk. Lab., 26: No. 7, 859 (1960).
25. N. S. Kurnakov, Introduction to Physicochemical Analysis, Izd. AN SSSR, 1940.
26. G. I. Taylor, Proc. Roy. Soc. (A), 145: 362 (1934).
27. E. Orowan, Z. Physik, 89: 634 (1934).
28. M. Polani, Z. Physik, 7: 323 (1921).
29. T. A. Kontorova and Ya. I. Frenkel', Zh. Eksperim. i Teor. Fiz., 89: 1348 (1938).
30. A. A. Griffith, Phil. Trans. Roy. Soc., 221: 163 (1921).
31. E. Orowan, Z. Physik, 82: 235 (1933).
32. A. F. Ioffe, M. V. Kiricheva, and M. A. Levitskaya, Z. Physik., 22: 286 (1924).
33. N. G. Petch, in: Progress in the Physics of Metals, No. 2, Metallurgizdat, 1958.
34. N. S. Kurnakov and S. A. Pogodin, Izv. Sektora Fiz.-Khim. Analiza Akad. Nauk SSSR, 16: No. 1, 7 (1943).
35. M. Potak and V. Sazonov, Zh. Tekhn. Fiz., No. 3 (1949).
36. J. M. Hodge, R. D. Monning, and H. M. Reichold, J. Metals (Metal Trans.), 185: 233 (1949).
37. Ch. Clark and A. E. White, Proc. Am. Soc. Inst. Mater., 32: 492 (1932).
38. A. P. Smiryagin, Industrial Nonferrous Alloys, Metallurgizdat, 1956.
39. A. P. Gulyaev, Metal Science, Oborongiz, 1951.

40. C. Z. Bokshtein, Structure and Mechanical Properties of Alloyed Steels, Metallurgizdat, 1954.

41. A. I. Gardin, Electron Microscopic Studies of Steels, Metallurgizdat, 1954.

42. M. P. Brau, Fracture and Brittleness of Structural Alloyed Steels [Russian translation], Mashgiz, 1960.

43. L. M. Utevskii, Dokl. Akad. Nauk SSSR, 119: No. 1 (1958).

44. H. G. Tapsell, The Creep of Metals, Oxford, 1931, p. 214.

45. R. N. Troitskii, Properties of Cast Iron, Metallurgizdat, 1938.

46. L. Landau and E. Livshits, Statistical Physics, Tekhteoretizdat, 1951.

47. Ya. I. Frenkel', Statistical Physics, Izd. AN SSSR, 1948.

48. B. Ya. Pines, Fiz. Zh., No. 3, 309 (1940).

49. I. L. Aptekar', Zh. Eksperim. i Teor. Fiz., 21: No. 8, 1951.

50. S. N. Zadumkin, Dokl. Akad. Nauk SSSR, 130: No. 4, 810 (1960).

51. R. Becker, Ann. Phys. (Paris), 32: 128 (1938).

52. A. S. Skapski, J. Chem. Phys., 16: 839 (1948).

53. K. A. Oriani, J. Chem. Phys., 18: 575 (1950).

54. V. K. Semenchenko, Surface Phenomena in Metals and Alloys, Tekhteoretizdat, 1957.

55. D. S. Kamenetskaya, Problems of the Physics of Metals and Metal Science, Metallurgizdat, 1949.

56. G. Borelius, Ann. Phys., 5: No. 98, 507 (1937).

57. W. Heitler, Ann. Phys., 4: No. 80, 1926.

58. K. F. Herfeld and W. Heitler, Z. Elektrochem., 31: 536 (1925).

59. B. Ya. Pines, Zh. Eksperim. i Teor. Fiz., 13: No. 11-12, 1943.

60. V. I. Danilov and D. S. Kamenetskaya, Zh. Fiz. Khim., 22: 69 (1948).

61. D. S. Kamenetskaya, Zh. Fiz. Khim., 22: 81 (1948).

62. B. Ya. Pines, Fiz. Zh., No. 3, 309 (1940).

63. B. Ya. Pines, Izv. Sektora Fiz.-Khim. Analiza Akad. Nauk SSSR, 16: 18 (1943).

64. B. Ya. Pines, Notes on the Physics of Metals, Izd. KhGU, 1961.

65. K. Wagner, Thermodynamics of Alloys [Russian translation], Metallurgizdat, 1957.

66. Shreder, Gorn. Zh., 4: 212 (1890).

67. Van Laar, Six Lectures on Thermodynamic Potential and Its Application to Chemical and Physical Equilibrium [Russian translation], ONTI, 1938.

68. Fischer, Z. Techn. Physik., 6: No. 4, 1925.

69. G. Tammann, Z. Anorg. Chem., 186: No. 3, 4 (1930).

70. M. V. Mal'tsev, in: Reports of MITsMiZ, No. 8, 1940, p. 67.

71. D. S. Kamenetskaya, Zh. Neorgan. Khim., 3: No. 3, 1958.

72. H. K. Hardy, Acta Met., 1: No. 2, 202 (1953).

73. W. Olsen, E. Schurmann, and G. Heynert, Arch. Eisenhüttenw., 26: No. 1, 29 (1955).

74. P. Gurevich, Fundamentals of Physical Kinetics, Tekhteoretizdat, 1940.

75. R. Berrer, Diffusion in Solids [Russian translation], IL, 1948.

76. S. D. Gertsriken and I. Ya. Dekhtyar, Diffusion in Metals and Alloys, Fizmatgiz, 1960.

77. Ya. I. Frenkel', Introduction to the Theory of Metals, Fizmatgiz, 1958.

78. C. Zener, Imperfections in Nearly Perfect Crystals, New York, 1952, p. 2.

79. A. D. Le Claire, Phil. Mag., 42: 673 (1951).

80. L. Landau,"Equilibrium Shapes of Crystals," in a Collection in honor of the 70th Birthday of A. F. Ioffe, Izd. AN SSSR, 1950.

81. D. Mak Lin, Grain Boundaires in Metals, Metallurgizdat, 1960.

82. S. D. Gertsriken and I. Ya. Dekhtyar, Vopr. Fiz. Metal. Metalloved., Akad. Nauk UkrSSR (1948), 135; (1950), 108.

83. M. E. Blanter, Zh. Tekhn. Fiz., 18: 529 (1948).

84. Ya. E. Geruzin, Usp. Fiz. Nauk, 61: No. 2, 217 (1957).

85. A. D. Le Claire and R. S. Barnes, J. Metals, 3: 1060 (1951).

86. S. T. Konobeevskii, Zh. Eksperim. i Teor. Fiz., 13: 418 (1943).

87. Holbrock, Fournas, Ind. Ing. Chem., 24: 993 (1932).

88. M. Paschke and Hauttman, Arch. Eisenhüttenw., 9:305 (1935).

89. D. W. Morgan and I. A. Kitchener, Trans. Faraday Soc. 50:51 (1954).

90. Crystal Growth, Discussions Faraday Soc., 5 (1949).
91. Bardenheier and Bleckmann, Eisenforsch., 21: 201 (1939).
92. D. S. Kamenetskaya, Problems of Metal Physics and Metal Science, Metallurgizdat, 1959.
93. C. E. Madenhall and L. R. Ingersol, Phil. Mag., 15:205 (1908).
94. V. I. Danilov and V. E. Neimark, Zh. Eksperim. i Teor. Fiz., 5: Nos. 9-10, 1937.
95. V. I. Schaefer, Science, 104: 455 (1946).
96. B. Vonnegut, J. Colloid Sci., 3: No. 6, 563 (1948).
97. D. Turnbull, J. Appl. Phys., 3: No. 6, 563 (1949).
98. H. Miers and F. Isaac, J. Chem. Soc., 89: 413 (1906).
99. R. E. Cech and D. J. Turnbull, J. Metals, 3, 242 (1951).
100. J. H. Hollomon and D. J. Turnbull, J. Metals, 3: 804 (1951).
101. M. Hillert, Acta Met., No. 6 (1951).
102. I. V. Mokhov and I. V. Kirsh, in: Critical Phenomena and Fluctuations in Solutions, Izd. AN SSSR, 1960.
103. H. K. Hardy and T. J. Hill, "Precipitation Processes," in: Progress in Metal Physics, No. 2 [Russian translation], Metallurgizdat, 1958.
104. Yu. A. Bagaryatskii, Zh. Tekhn. Fiz., 18: 827 (1948).
105. A. V. Shubikov, How Crystals Grow, Izd. AN SSSR, 1935.
106. V. D. Kuznetsov, Crystals and Crystallization, Tekhteoretizdat, 1953.
107. V. D. Kuznetsov, Physics of Solids, Vol. 1, Tomsk. Obl. Izd., 1937.
108. Bakli, Crystal Growth [Russian translation], IL, 1954.
109. D. D. Saratovkin, Dendritic Crystallization, Metallurgizdat, 1956.
110. A. Varma, Crystal Growth and Dislocations [Russian translation] IL, 1958.
111. B. Chalmers, in: Impurities and Defects, Metallurgizdat, 1960.
112. U. M. Martins, Progress in the Physics of Metals, Vol. 2, Metallurgizdat, 1958.
113. V. A. Indenbom, Kristallografiya, 3: No. 1, 113 (1958).
114. V. I. Danilov, Problems of Metal Science and Metal Physics, Metallurgizdat, 1949.
115. A. A. Bochvar, Mechanism and Kinetics of Crystallization of Eutectic Alloys, ONTI, 1935.
116. A. A. Bochvar and O. S. Zhadaeva, Izv. Akad. Nauk SSSR, Otd. Tekhn. Nauk, Nos. 4-5, 293 (1944).
117. R. Marc, Z. Physik. Chem., 67: 470 (1909).
118. J. Lebrun, Bull. Classe Sci. Acad. Roy. Belg, 970 (1913).
119. In: Impurities and Defects, Metallurgizdat, 1960.
120. A. A. Noges and W. R. Whitney, Z. Physik. Chem., 23: 689 (1897).
121. I. L. Mirkin, in: Reports of MIS, 1941.
122. W. Nernst, Z. Physik. Chem., 47: 52 (1904).
123. I. I. Andreev, Zh. Russ. Fiz.- Khim. Obshchestva, 40: No. 3, 397 (1908).
124. K. F. Bunchah and K. F. Mehl, Trans. AIME, 197: 1251 (1953).
125. G. L. Ivantsov, Dokl. Akad. Nauk SSSR, 58: 567 (1947).
126. B. Ya. Lyubov, Problems of Metal Science and Physics of Metals, Metallurgizdat, 1949.
127. B. Ya. Lyubov, Zh. Tekhn. Fiz., 20: 1344 (1950).
128. B. Ya. Lyubov and N. S. Fastov, Dokl. Akad. Nauk SSSR, 84: 939 (1952).
129. W. Brandt, J. Appl. Phys., 16: 139 (1948).
130. B. Ya. Pines, Zh. Eksperim. i Teor. Fiz., 18: 831 (1938).
131. B. Ya. Lyubov, Zh. Eksperim. i Teor. Fiz., 20: 872 (1959).
132. S. M. Danilov and V. I. Malkin, Problems of Metal Science and Physics of Metals, Metallurgizdat, 1951.
133. V. I. Malkin, Problems of Metal Science and Physics of Metals, Metallurgizdat, 1959.
134. Hull, Cotton and Mehl. Trans. AIME, 150: 185 (1942).
135. B. Ya. Lyubov, Problems of Metal Science and Physics of Metals, Metallurgizdat, 1952.
136. G. V. Kurdyumov and A. G. Khandros, Dokl. Akad. Nauk SSSR, 66: 211 (1949).
137. G. G. Lemlein, Dokl. Akad. Nauk SSSR, 34: No. 6, 1167 (1952).
138. I. V. Salli, Dokl. Akad. Nauk SSSR, 25: No. 1, 1953.
139. I. V. Salli, Zh. Eksperim. i Teor. Fiz., 25: No. 2, 1953.
140. M. P. Arbuzov, Vopr. Metal. i Fiz. Metalloved., Akad. Nauk UkrSSR, No. 3, 1952.
141. G. Edmunds, Trans. AIME, 143: 183 (1941).

142.   Born and Stern, Berlin Pruss. Acad., 901 (1909).

143.   E. N. Gapon, Zh. Russ. Fiz.-Khim. Obshchestva, 59: 933 (1927).

144.   G. V. Vul'f, Fundamentals of Crystallography, Gosizdat, 1927.

145.   P. Curie, Z. Krist., 12: 651 (1887).

146.   L. I. Kogan and V. E. Neimark, Problems of Metallography and Physics of Metals, Metallurgizdat, 1949.

147.   I. V. Salli and E. V. Finagina, Nauchn. Zap. Dnepropetr. Gos. Univ., 45, 1959.

148.   A. P. Gulyaev and E. V. Petunin, Tr. Tsentr. Nauchn.- Issled. Inst. Tekhnol. i Mashinostr., No. 5, Mashgiz, 1952.

149.   V. N. Gridnev and V. I. Trofimov, Dokl. Akad. Nauk. SSSR, 95: 741 (1954).

150.   L. Kaufman and M. Cohen, Trans. AIME, 206: 1393 (1956).

151.   L. Kaufman and M. Cohen, in: Progress in Physics of Metals, Vol. 4, Metallurgizdat, 1961.

152.   G. V. Kurdyumov, Reports of the Leningrad Conference on Industrial Applications of X-Rays, Mashgiz, 1949.

153.   I. N. Golikov, Dendritic Liquation of Steel, Metallurgizdat, 1958.

154.   K. P. Bunin, Iron-Carbon Alloys, Mashgiz, 1949.

155.   K. P. Bunin and Ya. N. Malinochka, Introduction to Metallography, Metallurgizdat, 1954.

156.   V. Peshkov, Zh. Fiz. Khim., No. 8, 835 (1946).

157.   Ya. V. Grechnyi, Dokl. Akad. Nauk SSSR, No. 1, 259 (1950).

158.   M. Hasselblatt, Z. Physik., 83 (1913).

159.   A. A. Popov, Problems of Metal Science and Heat Treatment, Mashgiz, 1956.

160.   Yu. A. Krishtal, Izv. Vuzov, Chernaya Metallurgiya, No. 3, 1960.

161.   V. M. Glazov and V. N. Vigdorovich, Dokl. Akad. Nauk SSSR, 118: No. 5, 924 (1958).

162.   N. N. Gorchakov, in: Cast and Fused-On Tools, Mashgiz, 1951.

163.   W. T. Olsen and R. Hultgren, J. Metals (Oct. 1950), p. 1323.

164.   E. Raub and A. Engel, Z. Electrochem., 49: 89 (1943).

165.   D. I. Lainer, in: The Use of X-Rays in the Study of Materials, Metallurgizdat, 1949.

166.   L. S. Palatnik and B. G. Boiko, Izv. Vuzov, Fizika, No. 8, 112 (1958).

167.   G. Falkengagen and W. Hofmann, Z. Metallk., 3: 43 (1954).

168.   I. N. Fridlyander, Dokl. Akad. Nauk SSSR, 104: No. 3 (1955).

169.   D. S. Kamenetskaya, Problems of Metal Science and Physics of Metals, Metallurgizdat, 1952.

170.   P. Duwer, R. H. Willens, and W.Klement, J. Appl. Phys., No. 3, 6, (1960).

171.   Heidenreich, Acta Met., 3: 79 (1955).

172.   F. Pawlek, Z. Metallk., 36: 105 (1944).

173.   I. S. Miroshnichenko and I. V. Salli, Izv. Vuzov, Chernaya Metallurgiya, No. 8 (1960).

174.   I. V. Salli and I. S. Miroshnichenko, Dokl. Akad. Nauk SSSR, 132: No. 6 (1960).

175.   I. S. Miroshnichenko, Izv. Vuzov, Tsvetnaya Metallurgiya, No. 1 (1961).

176.   I. S. Miroshnichenko and I. V. Salli, Izv. Akad. Nauk Otd. Tekhn. Nauk, SSSR, Met. i Toplivo, No. 3 (1961).

177.   I. S. Miroshnichenko and I. V. Salli, Zavodsk. Lab., No. 11 (1959).

178.   S. A. Saltykov, Stereometric Metallography, Metallurgizdat, 1958.

179.   K. P. Bunin, Metalloved. i Obrabotka Metall., No. 1 (1959).

180.   K. P. Bunin, Dokl. Akad. Nauk UkrSSR, No. 2 (1959).

181.   Roberts, J. Metals, 5: No. 2 (1959).

182.   G. Falkengagen and W. Hofmann, Arch. Eisenhüttenw., 23 (1952).

183.   Treatment of Metals, Handbook, Metallurgizdat, 1952.

184.   M. Khansen and K. Anderko, Structure of Nonferrous Alloys, Metallurgizdat, 1962.

185.   M. P. Slavinskii, Physicochemical Properties of Elements, Metallurgizdat, 1952.

186.   R. H. Schaefer and T. E. Kilgren, Metals Handbook ASM, 16, 1939, p. 24.

187.   H. Morrogh and W. Williams, Iron Steel Inst., 155: 321 (1947).

188.   A. P. Gulyaev and E. F. Trusova, Zh. Tekhn. Fiz., 20: No. 1 (1950).

189.   W. Koster and E. Schmid, Z. Metallk., 19: 232 (1937).

190.   B.G. Livshits, Physical Properties of Metals and Alloys, Mashgiz, 1956.

191.   M. A. Krishtal, Fiz. Metal. i Metalloved., 9: No. 5 (1960).

192. A. G. Lesnik, Models of Atomic Interactions in the Statistical Theory of Alloys, Fizmatgiz, 1962.
193. I. V. Salli, Dokl. Akad. Nauk UkrSSR, No. 5 (1951).
194. V. I. Psarev and I. V. Salli, Fiz. Metal. i Metalloved., 5: No. 2 (1957).
195. V. Smolukhovskii, Coagulation of Colloids, ONTI NKTP SSSR, 1936.
196. I. L. Mirkin, Tr. Inst. Stali, 1: No. 18, 73 (1941).
197. N. Beljuew, Rev. Met. (Paris), No. 12 (1930).
198. Schimyra and Esser, Stahl u. Eisen, 48: 1674 (1930).
199. O. Green, Trans. Am. Sec. Steel Treating, 21: 57 (1929).
200. S. Z. Bokshtein, Zh. Tekhn. Fiz., 22: No. 12 (1947).
201. Gensamer, Trans. Am. Soc. Metals.
202. C. S. Ljalikov, Acta Physicochim. URSS (1940), 1243.
203. S. V. Chardyntsev, Tr. GOI, 9: No. 8 (1933).
204. S. T. Konobeevskii, Izv. Akad. Nauk SSSR, Ser. Khim., No. 5, 1202 (1937).
205. N. N. Sirota, Dokl. Akad. Nauk SSSR, 50: No. 6 (1946).
206. O. M. Todes, Zh. Fiz. Khim., 20: No. 2 (1946).
207. I. M. Livshits and V. V. Slezov, Zh. Eksperim. i Teor. Fiz., 35: 474 (1958).
208. V. I. Psarev, Dissertation, Dnepropetrovskii Metallurgicheskii Institut, 1956.
209. A. I. Dukhin, Problems of Metal Science and Physics of Metals, Metallurgizdat, 1959.
210. V. I. Danilov and Yu. A. Krishtal, Problems of Metal Science and Physics of Metals, Metallurgizdat, 1949.
211. L. I. Kogan, V. E. Neimark, I. B. Piletskaya, and R. I. Entin, Problems of Metals Science and Physics of Metals, Metallurgizdat, 1949.
212. F. C. Thompson, Trans. Faraday Soc., 17: 391 (1922).
213. C. Zener, Metals. Technol., 13: 1925 (1946).
214. K. T. Aust and B. Chalmer, Proc. Roy. Soc., 204: 210 (1950).
215. K. T. Aust and B. Chalmer, Proc. Roy. Soc., 204: 354 (1950).
216. I. V. Salli, Dokl. Akad. Nauk UkrSSR, No. 5 (1950).
217. M. A. Krishtal, Fiz. Metal. i Metalloved., 2: No. 2 (1956).
218. I. V. Salli, Dokl. Akad. Nauk UkrSSR, No. 4 (1952).
219. V. I. Psarev and I. V. Salli, Nauchn. Zap. Dnepropetr. Gos Univ., 45 (1956).
220. V. I. Psarev, Izv. Vuzov, No. 2 (1959).
221. I. V. Salli, Izv. Vuzov, No. 5 (1960).
222. I. V. Salli, Fiz. Metal. i Metalloved., 8: No. 5 (1959).
223. P. J. Cleman and J. C. Fischer, Acta Met., 3: No. 1 (1955).
224. D. Turnbull, J. Chem. Phys., 20: 411 (1952).
225. D. M. Pound and V. R. La Mar, J. Am. Chem. Soc., 74: 23 (1952).
226. D. J. Turnbull and R. E. Cech, J. Appl. Phys., 21: 840 (1950).
227. C. E. Mandenhall and L. R. Ingersoll, Phil. Mag., 15: 205 (1908).
228. V. J. Schaefer, Ind. Eng. Chem., 44: 1300 (1952).
229. K. I. Vashchenko, Modified Cast Iron, Mashgiz, 1946.
230. N. G. Girshovich, Casting of Cast Iron, Mashgiz, 1949.
231. K. P. Bunin, G. I. Ivantsov, and Ya. N. Malinochka, Structure of Cast Iron, Mashgiz, 1952.
232. I. N. Bogachev, Metallography of Cast Iron, Metallurgizdat, 1952.
233. K. P. Bunin, Whitened Cast Iron, Metallurgizdat, 1945.
234. V. T. Zubarev, Theoretical Foundations of Graphitization of White Cast Iron and Steel, Mashgiz, 1957.
235. K. P. Bunin, A. A. Baranov, and É. N. Pogrebnoi, Graphitization of Steel, Izd. AN UkrSSR, 1961.
236. I. V. Salli, Izv. Akad. Nauk SSSR, Otd. Techn. Nauk, No. 12 (1956).
237. I. V. Salli, Izv. Vuzov, Chernaya Metallurgiya, No. 10 (1961).
238. I. N. Bogachev and L. D. Lobachev, Metallurg. No. 7-8 (1938).
239. K. P. Bunin and D. Katsnel'son, Metallurg, No. 8 (1939).
240. E. Chimenguej and R. Ensminger, Trans. AIME, 67: 392 (1922).
241. L. I. Shushpanov, Metallurg, No. 6 (1937).
242. I. Smith and M. Olney, Research, 3 (Apr. 1950).
243. M. G. Lozinskii, Vacuum Metallography, Mashgiz, 1956.

244. I. V. Salli and A. N. Shul'diner, Zh. Tekhn. Fiz., 23: No. 2 (1953).

245. I. V. Salli and E. Z. Graifer, Dokl. Akad. Nauk SSSR, 97: No. 4 (1954).

246. I. I. Pyasetskii and I. V. Salli, Fiz. Metal. i Metalloved., 3: No. 3 (1956).

247. I. I. Pyasetskii and I. V. Salli, Zavodsk. Lab., No. 2 (1955).

248. I. I. Pyasetskii, Dokl. Akad. Nauk UkrSSR, No. 2 (1958).

249. V. N. L'nyanoi and I. V. Salli, Nauchn. Dokl. Vysshei. Shkoly, No. 3 (1958).

250. V. N. L'nyanoi, Izv. Akad. Nauk SSSR, Otd. Tekhn. Nauk, No. 8 (1957).

251. V. N. L'nyanoi, Nauchn. Zap. Dnepropetr. Gos. Univ., 72: No. 8 (1957).

252. V. N. L'nyanoi and I. V. Salli, Izv. Vuzov, Chernaya Metallurgiya, No. 3 (1960).

253. I. N. Stranski, Z. Phys. Chem., 17: 127 (1932).

254. A. A. Baikov, Collection of Reports, Vol. 2, Izd. AN SSSR, 1948.

255. S.M. Palestin, Metallurg, No. 4 (1938).

256. K. P. Bunin and I. V. Salli, Dokl. Akad. Nauk SSSR, 83: No. 6 (1952).

257. L. S. Darken, Trans ASME, 175: 184 (1948).

258. G. S. Hartlay and I. Sranki, Trans. Farad. Soc., 45: 801 (1949).

259. N. N. Sirota, Dokl. Akad. Nauk SSSR, 51: 295 (1946).

260. A. S. Palatnik and V. S. Zorin, Zh. Fiz. Khim., 33: No. 8 (1959).

261. V. N. L'nyanoi and I. V. Salli, Fiz. Metal. i Metalloved., 9: No. 3 (1960).

262. I. M. Pronman, Liteinoe Proizv., No. 4 (1960).

263. I. M. Pronman, Liteinoe Proizv., No. 11 (1960).

264. F. G. Frank, Discussions Farad. Soc., 5: 48 (1949).

265. K. P. Bunin and I. M. Danil'chenko, Dokl. Akad. Nauk UkrSSR, No. 4 (1950).

266. B. Yu. Mett and R. I. Éntin, Dokl. Akad Nauk SSSR, 68: No. 4 (1949).

267. M. Cohen, E. S. Machlin, and V. G. Paranype, Thermodynamics in Physical Metallurgy, ASM, 1949.

268. E. S. Machlin, Trans. AIME, 200: 284 (1954).

269. J. W. Gibbs, Trans. Can. Acad., 3: 108, 343, (1875-78).

270. D. K. Chernov, "Investigation of the Structure of Cast Steel Ingots," A report to the I. R. T. Society, December 2, 1878.

271. G. Tammann, Z. Phys. Chem., 23: 326 (1897).

272. Kossel, Naturwiss., 18: No. 9 (1930).

273. M. Folmer, Die Kinetik der Phasenbildung, 119: 277 (1921).

274. J. N. Stranskj and R. Kaischew, Z. Anorg. Allgem. Chem., 26: 312 (1934).

275. E. C. Ellwood, J. Inst. Metals, 80: No. 5, 217 (1951-1952).

276. A. I. Kolmogorov, Izv. Akad. Nauk SSSR, No. 3, 353 (1937).

277. B. Koenigmann, Growth and Shape of Crystals [Russian translation], IL, 1961.

278. I. N. Stranski, Z. Phys. Chem., 17: 127 (1932).

279. R. Fogel, Z. Anorg. Allgem. Chem., 116: 21 (1921).

280. A. Papapetrou, Z. Krist. 92: 89 (1935).

281. G. W. Sears, Metal Progress, No. 1 (1956).

282. S. S. Brenner and G. W. Sears, Acta Met., 4: 268 (1956).

283. I. A. Oding, Izv. Akad. Nauk SSSR, Otd. Tekhn. Nauk, No. 12 (1948).

284. A. F. Ioffe, Physics of Crystals, Gosizdat, 1929.

285. W. Reed, Dislocations in Crystals [Russian translation], IL, 1957.

286. B. Chalmers, Proc. Roy. Soc., 175: 100 (1940).

287. G. Chandron, P. Lacomb, and N. Gannaguist, Nature, 162: 854 (1948).

288. W. I. Pumphrey and L. Lions, J. Inst. Metals, 89: 33 (1953).

289. D. M. Lachlan, Acta Met., 5: No. 2, 111 (1957).

290. S. Glasston, K. Leidler and H. Eyring, Theory of Absolute Reaction Rates [Russian translation], IL, 1948.

291. A. E. Stern, E. M. Irich, and H. Eyring, J. Phys. Chem., 44: 981 (1940).

292. A. V. Belyustin, Zh. Eksperim. Teor. Fiz., 28: No. 15 (1955).

293. I. I. Kornilov, Physicochemical Foundations of the Heat Resistance of Metals and Alloys, Izd. AN SSSR, 1962.

294. M. I. Shakparonov, Introduction to the Molecular Theory of Solutions, Gostekhizdat, 1956.

295. M. I. Shakhparonov, in: Structure and Properties of Liquid Metals, Izd. Inst. Metallurgii, Moscow, 1959.

296. N. S. Kurnakov, Selected Works, Vols. 1, 2, Izd. AN SSSR, 1960, 1961.

297. A. A. Bochvar, Izv. Akad. Nauk SSSR, Otd. Tekhn. Nauk, No. 10 (1947).

298. A. A. Bochvar, Metal Technology, Collection of Reports of Mintsvetmetzoloto, No. 18 (1947).

299. A. A. Bochvar and O. S. Zhadaeva, Izv. Akad. Nauk SSSR, Otd. Tekhn. Nauk. No. 10-11 (1945).

300. V. M. Glazov and V. N. Vigdorovich, Microhardness of Metals, Metallurgizdat, 1962.

301. A. A. Bochvar, Metal Science, Metallurgizdat, 1956.

302. A. A. Bochvar, Fundamentals of Heat Treatment, ONTI, 1931.

303. V. M. Glazov and Lyu Chzhen'-yuan', Izv. Akad. Nauk SSSR, Otd. Tekhn. Nauk Met. i Toplivo, No. 2 (1961).

304. D. S. Kamenetskaya, É. V. Rakhmanova, and E. Z. Spektor, Dokl. Akad. Nauk SSSR, 142: No. 3 (1962).